Tools and Techniques in Radiation Biophysics

Ashima Pathak

Tools and Techniques in Radiation Biophysics

Springer

Ashima Pathak
Department of Biotechnology
Goswami Ganesh Dutta Sanatan Dharma College
Chandigarh, India

ISBN 978-981-99-6088-0 ISBN 978-981-99-6086-6 (eBook)
https://doi.org/10.1007/978-981-99-6086-6

This Springer imprint is published by the registered company Springer Nature Singapore Pte Ltd.
The registered company address is: 152 Beach Road, #21-01/04 Gateway East, Singapore 189721, Singapore

Preface

This book is a result of insights got while teaching a course on radiation biophysics. In choosing the course text, it was necessary to propose to the students a large number of references, only part of these would ultimately be used in the course. My aim has been to produce a text that would be useful for advanced undergraduates and post-graduate students in the appropriate sciences while at the same time filling a need for a desk-top reference for working scientists in this field. I have tried to keep the mathematical derivations to a bare minimum so that even a biology student could easily grasp the concepts.

I have a great pleasure in presenting the first edition of a book on *Tools and Techniques in Radiation Biophysics* with several unique/novel features. Since there have been tremendous advancements in the field of radiation biophysics and because of the lack of a comprehensive material available at one platform that can help the students understand the important concepts, I was prompted to write this book incorporating all the important concepts from their basics to the applications in an easy to grasp language.

Although emphasis of the book is on the basic concepts of radiation biophysics, discussion is also undertaken to show how these concepts help in understanding the diagnosis and therapy of various diseases. The motive is to provide a clear, accurate, and up-to-date introduction to radiation biophysics for graduate, postgraduate students, researchers, and clinicians of various disciplines and in doing so, each element of this book presents a learning opportunity.

This book progresses from the physical interactions to the radiochemical interactions to the biological sequelae. It begins with a short historical background to the field in Chapter 1. Chapter 2 covers the basic concepts of the atomic and nuclear structure. Chapter 3 deals with the introduction to the radioactivity and various units that are used to express the radioactivity under different circumstances. Chapter 4 includes different radioactive decay laws involving all three kinds of radiation and how they interact with the matter through which they pass is dealt with in Chapter 5. Chapter 6 focusses on the different types of radiation present in the environment and how they affect living beings. Chapter 7 emphasises on the measurement of different radiations on the basis of different principles using different techniques. How these radionuclides are involved in diagnosis (Chapter 8) and therapy (Chapter 9) are also addressed in detail. Chapter 10 discusses about the impact or efficiency of radiations on cancer—diagnosis as well as treatment. Emphasis is also given to various applications of radionuclides in metabolic and biological effects (Chapter 11) and in molecular biology (Chapter 12) and also other applications in various fields of life (Chapter 13). Finally, the dosimetry and safety issues regarding radiation are also covered in detail in Chapter 14. The order of the chapters is kept in such a manner that helps in slow buildup of the basic concepts of the subject to its advanced applications.

Graphs, diagrams, and flow charts have been incorporated to make the more instructive, meaningful, and comprehensible. The content of the book is made as concise as possible to enhance the clarity so that the book is helpful and useful to all the readers. The text is a pleasure to read, and topics are developed logically with all the latest advances and developments. Difficult concepts are explained clearly supported with plentiful illustrations. It is an introductory text that begins at the beginning and does not assume that the reader has any prior knowledge of the concepts and the technique used. So the concepts are explained to make it more approachable for students who are new to this field.

A set of learning objectives are added at the beginning of each chapter that provide a road map through the chapter and also enable the reader understand what they should be able to do with the material in the chapter once it has been mastered and also these objectives direct the students to apply the concepts contained in the chapter rather than memorise facts. Each chapter begins with an overview of the importance of its contents and concludes with a summary reviewing the major topics covered. The key concepts are put in focus boxes to facilitate learning and grasping important findings about a particular topic without distracting from the flow of information. Each chapter also includes the check points at the end to test the student's ability to apply the principles to solving problems rather than test basic fact-based recall. This will also give students the opportunity to practice what they have learned.

Readers will appreciate the inclusion of glossary of text terms at the end. A short list of review articles, monographs, book chapters, or websites is also provided where the readers can obtain additional and relevant information associated with each chapter.

It is true that I represent a selected group of individuals authoring books, having some time at disposal, besides hard work, determination, and dedication. I consider myself as a regular student of radiation biophysics. However, it is beyond my capability to keep track of the ever-growing advances in this field due to the exponential growth of the subject. I honestly admit that I have to depend on mature readers and authors to write this book.

The result is, I hope, a more balanced approach right from the basic concepts to all the latest advancements concerning numerous applications of radionuclides, with special mention to the safety issues.

Chandigarh, India Ashima Pathak

Acknowledgements

I would like to thank my husband Rajiv for his unending support. He played an important role in the planning and actualisation of the book. He also has a major hand in providing insightful ideas, constructive and invaluable suggestions that have helped me in giving a final shape to this work. I owe special thanks to my loving daughter Sakshi in providing all inputs and skilled knowledge regarding the designing of the large amount of artwork and also providing better and clearer illustrations that were necessary for this book. Her professional knowledge and guidance have made this book more impressive and presentable. I am also thankful to my parents who have always been an inspiration behind this venture, especially my father who helped me in the final editing of the book with his technical knowledge and professional acumen and suggestions—despite being the layman to this subject. My gratitude to him for keeping patience in going through the whole depth. Last but not least, I would like to thank my loving son Vibhu for his constant encouragement and for his moral support during my hard times.

I also acknowledge that this book represents the work not only of its author but also of all those whose experiments and writings provided me with ideas, inspiration, and information. I hope the final product reflects the high quality of the input from all the sources.

Finally, I express my deepest appreciation to the publisher Springer Nature that provided me the platform to make this project a reality. I am truly grateful to Ms. Stephanie, Project Coordinator (Books) for her constant support and assistance throughout the publishing process that has helped in making this venture a success.

About the Book

This edition emphasises the descriptive details of the study of radiations covering the basic concepts and their advanced applications. It also highlights essential and relevant content to clear the basics for students/researchers about the handling of radioisotopes and radiation measurements using various instruments. Further, it also focuses on the effects and up-to-date applications of radiation on biological systems and their uses in the diagnosis and therapy of various diseases. It also includes a special mention of the safety issues that must be taken into consideration while dealing with radioisotopes.

The book is organised into various chapters that systematically take the reader from the very basic concepts to the advanced knowledge of radiations. The content has been arranged in a way that it provides an easy understanding of the subject matter with the help of self-explanatory, well-illustrated figures and easy-to-grasp language along with key points and innovative questions in each chapter. This will enable the students to grasp the concepts efficiently. Thorough research was done on various topics before putting them into words.

This book is designed for undergraduate and post-graduate students of various disciplines to facilitate learning and deep understanding of the concepts involved in this area. Therefore, this book has been written with an aim to help the readers develop an interest in the subject so that they can pursue their careers efficiently in this field. Students, researchers, and clinicians can also use this book as an easy reference to clear their queries.

Contents

About the Author

Ashima Pathak Assistant Professor in the Department of Biotechnology, GGDSD College, Chandigarh. Her area of specialisation is radiation biophysics, biophysical techniques, and trace metal toxicology. Previously, she has also served in the Department of Nuclear Medicine, PGIMER, Chandigarh where she was involved in the handling of radioisotopes and various diagnostic machines like Gamma Camera, radiation dosimeters that are used in the clinical diagnosis and therapy of various diseases This practical knowledge has been incorporated and well explained in this book.

She did her graduation, post-graduation, and doctorate from the Department of Biophysics, Panjab University, Chandigarh. She was awarded a gold medal on graduation. She has guided a number of postgraduate students for their research work. She has been awarded UGC-funded grants and fellowships. She has published several research papers in various national and international reputed journals.

She is an author of the book *Fundamentals of Microscope.* Besides, she has also written two chapters in the book *Environment, Road Safety Education and Violence Against Women and Children.*

List of Abbreviations

2DE	Two-dimensional gel electrophoresis
2DEMS	Two-dimensional electrophoresis mass spectrometry
3D	Three-dimensional
3DCRT	Three-dimensional conformal radiation therapy
5-FU	Fluorouracil
ADME	Absorption, distribution, metabolism, and excretion
AFLP	Amplified fragment length polymorphism
Ag	Antigen
ALARA	As low as reasonably achievable
Ar	Argon
ATP	Adenosine triphosphate
B	Boron
BaF_2	Barium fluoride
bisMSB	p-Bis (0-methyl-styryl)-benzene]
BNCT	Boron neutron capture therapy
Butyl PBT	Butyl polybutylene terephthalate
C	Carbon
Ca	Calcium
CaF_2	Calcium fluoride
cDNA	Complementary DNA
CLIA	Chemiluminescence immunoassay
CNS	Central nervous system
CO_2	Carbon dioxide
COOH	Carboxyl group
Cs	Cesium
CsI	Cesium iodide
CT	Computed tomography
DGGE	Denaturing gradient gel electrophoresis
DMS	Dimethyl sulfate
DNA	Deoxyribonucleic acid
DNase	Deoxyribonuclease

e	Electron
EB	Binding energy
ECT	Emission computed tomography
E_e	Kinetic energy
ELISA	Enzyme-linked immunosorbent assay
EPD	Electronic personal dosimeter
Eu	Europium
$E\gamma$	Gamma ray energy
FAD	Flavin adenine dinucleotide
FDA	Food and Drug Administration
FDG	Fluorodeoxyglucose
FDG-PET	Fluorodeoxyglucose positron emission tomography
Fe^{2+}	Ferrous ion
Fe^{3+}	Ferric ion
GCMS	Gas chromatography-mass spectrometry
Ge	Germanium
GM	Geiger–Müller
H_2O	Water
H_2O_2	Hydrogen peroxide
HCV	Hepatitis C virus
HDR	High dose rate
He	Helium
HIV	Human immunodeficiency virus
HPLC	High performance liquid chromatography
HTA	Heteroduplex tracking assay
HVL	Half value layer
I	Iodine
ICAT	Isotope-coded affinity-tag
IGRT	Image guided radiation therapy
IMRT	Intensity modulated radiation therapy
IORT	Intraoperative radiation therapy
IRMA	Immunoradiometric assay
Kr	Krypton
LC	Liquid chromatography
LCMS	Liquid chromatography mass spectrometer
LET	Linear energy transfer
Li	Lithium
LLD	Lower level discriminator
LSC	Liquid scintillation counter
mAb	Monoclonal antibody
MCA	Multi-channel analyser
MOSFET	Metal-oxide-semiconductor field-effect transistor
MRI	Medical resonance imaging
MS/MS	Tandem mass spectrometer
N	Nitrogen

Na	Sodium
NH_3	Ammonia
NH_4Br	Ammonium bromide
NO_2	Nitrogen dioxide
O	Oxygen
OSL	Optically stimulated luminescence
Pb	Lead
PCR	Polymerase chain reaction
PET	Positron emission tomography
PHA	Pulse height analyser
PM tubes	Photomultiplier tube
PMT	Photomultiplier tube
POPOP	1,4-(di-2-(5-phenyl oxazolyl))-benzoyl)
PPi	Pyrophosphate
PPO	2,5-biphenyl oxazole
PRRT	Peptide receptor radionuclide therapy
PS	Polystyrene
PVT	Polyvinyl toluene
QTL	Quantitative trait locus
RFLP	Restriction fragment length polymorphism
RIA	Radio immuno assay
RNA	Ribonucleic acid
RNase	Ribonuclease
SCA	Single channel analyser
SEM	Scanning electron microscope
Si	Silicon
Si	Système International
SIO_2	Silicon dioxide
SO_2	Sulfur dioxide
SPA	Scintillation proximity assay
SPECT	Single-photon emission computed tomography
SPRD	Self-reading pocket dosimeter
SSCP	Single stranded conformational polymorphism
TAT	Targeted alpha therapy
TB	Tuberculosis
TCA	Trichloro acetic acid
TEM	Transmission electron microscope
Tl	Thalium
TLD	Thermoluminescent dosimeter
ULD	Upper level discriminator
UV	Ultraviolet
ZnS	Zinc sulfide

Historical Perspectives of Radiation

Key Learning Objectives
- Radiobiology and key events
- The invention of X-rays
- Discovery of radioactivity
- Contribution of Marie Curie
- Rutherford's experiments
- Thomson's discovery
- Chadwick's experiment

Overview
Since the major discoveries by WC Roentgen, H Becquerel, Curies, E Rutherford, JJ, Thomson, and J Chadwick in the nineteenth century, a lot of studies have been done on different aspects of radiation. Because of all this research and experimentation and the advent of related sophisticated instruments, we are now able to diagnose and even treat various diseases using different types of radioactive materials.

1 Radiobiology

Radiobiology is concerned with how radiation affects and interacts with living things and systems. Radiation is the energy that moves through space or matter as a particle or a wave. Radiobiology is actually a combination of biology, physics, and epidemiology.

A. Pathak, *Tools and Techniques in Radiation Biophysics*,
https://doi.org/10.1007/978-981-99-6086-6_1

Key Concept
Radiation is the energy that emanates from a source and travels at the speed of light through space.

Three important events mark the beginning of radiobiology:

- The development of X-rays by Wilhelm Conrad Roentgen in 1895;
- The discovery of the rays emitted by a material containing uranium by Antoine Henri Becquerel in 1896; and
- In 1898, Pierre and Marie Curie discovered radium.

Since then, these radiations have played a crucial role in the study of atomic, nuclear, and medical physics.

2 Invention of X-rays

Key Concept
X-rays were discovered and named by WC Roentgen while working with fluorescent light bulbs.

In 1895, a German physicist, Roentgen, was using a cathode ray tube in his laboratory. He was using tubes that resembled modern day fluorescent lamps. He injected a special gas into the tube and passed a high voltage through it. He observed that the tube emitted a fluorescent radiance. In his subsequent experiments, he covered the tube with thick black paper and discovered that a green fluorescent light appeared on a screen that was placed a few feet from the tube. He came to understand that he had invented a "ray of invisible light" that could penetrate thick paper. The new invisible ray was given the name *X-ray*.

3 Discovery of Becquerel

French physicist Henri Becquerel was a professor of the third generation. His father, Alexander Edmond Becquerel was a professor of applied physics who conducted solar radiation and phosphorescence research. The electrolytic process of extracting metals from their ores was created by his grandfather, Antoine César Becquerel. Becquerel had helped his father with several experiments on phosphorescence. He continued his father's research into *phosphorescence*, the phenomenon in which certain substances emit visible light when exposed to a bright light source. The initial work of Henri Becquerel was focused on the plane polarisation of light, phosphorescence, and the crystals' light absorption. In 1896, Henri Becquerel used naturally fluorescent materials to investigate the properties of X-rays, which Wilhelm Roentgen had discovered in 1895. In 1896, a French physicist made the discovery of radioactivity while working with a photographic plate. At the age of 44, Becquerel was experimenting with a substance that could emit light when exposed to sunlight. A small amount of uranium salt was kept in one of his desk drawers. He believed that the uranium absorbed the sun's energy and then emitted it in the form of X-rays, so potassium uranyl sulphate was exposed to sunlight before being placed it on photographic plates that were wrapped in black paper. These photographic plates were then used to detect the light (known as *fluorescence*) emanating from the exposed samples. Becquerel was unable to conduct his experiments involving fluorescence induced by sunlight during a spell of overcast weather in Paris. Instead, he made the decision to develop some of his unexposed photographic plates in order to see any possible light leaks. He was astonished to see that the images were strong and distinct, i.e., he noticed that a plate was black when placed on uranium salt. Becquerel was initially puzzled, but then he realised that some unknown type of radiation was released from the uranium salt, and this happened without the aid of any external energy source such as the sun, hence the discovery of *radioactivity.* He shared the Nobel Prize in Physics with the Curies in 1903 for this discovery. The intensity unit of the radioactive source was named after him. 1 becquerel (abbreviated Bq) represents the average rate of disintegration of an atom per second in the source.

4 Curies' Contribution

Key Concept
Marie Curie was a Polish and naturalised French physicist and chemist who initiated research on radioactivity.

Another important landmark was when Marie Sklodowska-Curie and her husband, Pierre Curie, were able to isolate the radioactive elements from the parent rocks. The Curies extracted uranium from the ore and, to their astonishment, discovered that the leftover ore had higher activity than the uranium itself. They determined that the ore contained supplementary radioactive elements. The two radioactive elements were discovered and isolated after extensive efforts. The first was named *polonium* (after Marie's native Poland), and the second one was named *radium* (which means "the radiating substance"). The isolation of sufficient quantities of different elements so as to determine their chemical properties required an additional four years of processing tons of ores. Marie Curie died in 1934 from a blood disease, possibly leukaemia, that her work may have caused. Marie Curie coined the term radioactivity when she and her husband began investigating the phenomenon recently discovered by Becquerel.

Key Concept
Sklokkwsko-Curie was awarded the Nobel Prize in Chemistry in 1911 for discovering radium and polonium and isolating radium.

5 Rutherford Experiment

Key Concept
E. Rutherford is the founder of nuclear chemistry and physics. The atomic nucleus, the proton, the alpha particle, and the beta particle were discovered and named by him.

Ernest Rutherford, another scientist, began researching the impact of X-rays on various materials. Soon after the discovery of radioactivity, he started studying the particles emitted by uranium and its compounds. Later, he determined that the radiation he had detected could not have been X-rays, as X-rays are neutral and cannot be deflected by a magnetic field. However, the new radiation was deflected by the magnetic field, indicating that it is charged and distinct from the X-rays. When various radioactive substances were placed in a magnetic field, they get deflected in different directions or did not deflect at all, thus demonstrating that there were three classes of radioactivity: negative, positive, and electrically neutral. Ernest Rutherford, who conducted numerous experiments examining the properties of radioactive decay, classified and named these radiations as alpha, beta, and gamma particles and classified them by their ability to penetrate matter. He studied the absorption of radioactivity using thin sheets of metal and identified two components: alpha (α) radiation, which is absorbed by a very thin metal, and beta (β) radiation, which can pass through 100 times as much foil before being absorbed. Shortly thereafter, a third type of radiation, called gamma (γ) rays, was identified that can penetrate several centimeters of lead.

Rutherford used an apparatus similar to that depicted in Fig. 1.

When the chamber's air was removed, the alpha source left a mark on the photographic plate. With the introduction of air, the spot vanished. Thus, only a few centimeters of air were necessary to block the alpha radiation. Since alpha particles have a greater electric charge, are heavier, and move more slowly than beta or gamma particles, they interact with matter much more readily. Beta particles are still electrically charged despite being much lighter and moving faster than alpha particles. Gamma rays, however, carry no electric charge, so they can travel through materials for great distances without interacting. Several centimeters of lead or one meter of concrete are needed to block the majority of gamma rays. All these types of radiation and their properties will be examined in greater depth in the following chapters.

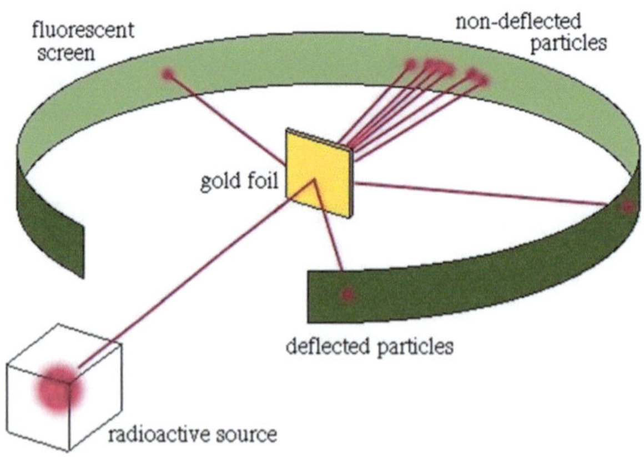

Fig. 1 Rutherford apparatus. http://large.stanford.edu/courses/2017/ph241/sivulka2/images/f1big.jpg. Adapted from Experimental Evidence for the *Structure* of the Atom by George Sivulka 2017

6 Thomson's Experiment

Inspired by the discovery of X-rays, the British physicist Sir Joseph John Thomson started his research on cathode rays and ultimately discovered the electron in 1896, the first sub-atomic particle. His discovery was the result of an attempt to resolve a long-standing debate regarding the nature of cathode rays, which are produced when an electric current is passed through a vessel from which most air and other gases have been evacuated. Thomson utilised an enhanced vacuum technique in the vessel to argue that these rays consisted of particles. This discovery of the electron has revolutionised our knowledge of atomic structure. He also proposed the plum pudding model of the atom, in which the negatively charged electrons were embedded in a positively charged "soup".

7 Chadwick's Discovery

J.J. Thomson might be described as "the man who first split the atom," and he was awarded the Nobel prize in 1906 and knighted in 1908. This also led to the development of the mass spectrograph. Prior to the outbreak of World War I, Thomson made another ground-breaking discovery: the isotope.

In 1932, the physicist, James Chadwick demonstrated the existence of neutrons, the electrically neutral elementary particles. This was a fundamental discovery in the field of nuclear science. He conducted the experiment by bombarding beryllium atoms with alpha particles from the natural radioactive decay of polonium. During this experiment, unidentified radiation was emitted. Chadwick interpreted this radiation as being composed of particles with a neutral charge and roughly the same mass as a proton-neutron. By the 1920s, the working model of the atom contained protons and electrons; Chadwick's work proved that neutrons also existed within the nucleus, changing the accepted atomic model. The Nobel Prize in Physics was awarded to him in 1935. All these contributions are summarised in the figure below (Fig. 2).

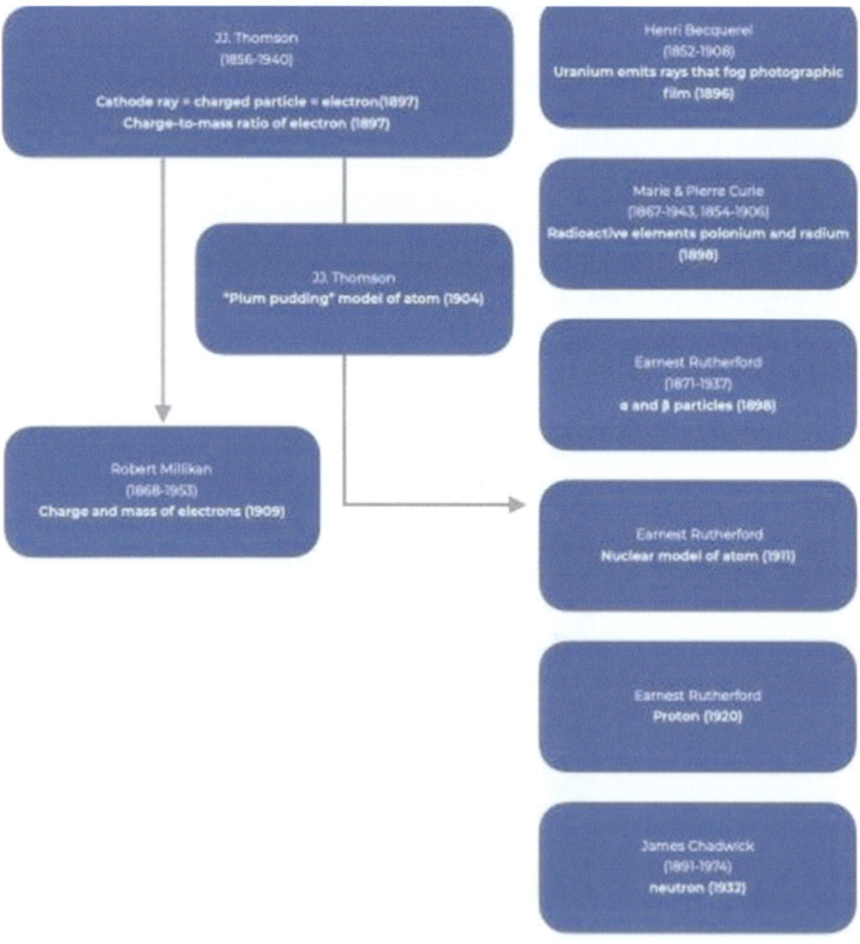

Fig. 2 Major contributions

Check Points

- What are the key events in the field of radiobiology?
- How did WC Roentgen discover X-rays?
- Describe the experimentation of Becquerel that has led to the discovery of radioactivity.
- How did the Curies contribute to the study of radioactivity?
- Explain Rutherford's experiments to invent different types of radiation.
- How did the work on cathode rays by J.J. Thomson lead to the discovery of the electron?
- How did Chadwick's experiments change the atomic model?

In summary, it is worth mentioning that the period from 1895 to 1905 was considered the "golden era" of radiation discovery and radioactivity research. These discoveries provided insight into studying the details of these radiations, such as the effect of radiations on biological systems and understanding their mechanisms of action. Later, the scientists started conducting experiments at the tissue level and then at the cellular level. Though earlier it was difficult to quantify these radiations, extensive studies led to the development of newer techniques and several sophisticated machines that are now being used to diagnose and cure several diseases. Nowadays, these radiations are being used in almost all fields. The latest progress in molecular biology and DNA research has provided new tools for the examination of damage and repair processes with the help of these radiations.

Further Reading

Garcia F, Arruda-Neto JDT, Manso MV, Helene OM, Vanin VR, Rodriguez O, Mesa J, Likhachev VP, Pereira Filho JW, Deppman A (1999) A new statistical method for transfer coefficient calculations in the framework of the general multiple- compartment model of transport for radionuclides in biological systems. Phys Med Biol 44 (10): 2463. https://doi.org/10.1088/0031-9155/44/10/308

Geiger H (1908) On the scattering of the α-particles by matter. Proceedings of the Royal Society A. Mathematical, Physical and Engineering Sciences. 81 (546): 174-177. https://doi.org/10.1098/rspa.1908.0067. ISSN (Print) 0950-1207; eISSN2053-9150

Gegier H, Marsden E (1909) On a diffuse reflection of the α-particles. Proceedings of the Royal Society A. Mathematical, Physical and engineering Sciences 81: 174. https://doi.org/10.1098/rspa.1909.0054. ISSN0950-1207. eISSN2053-9150.

Gegier H (1910) The scattering of the α-particles by matter. Proceedings of the Royal Society Series A 83 (565): 492-504.

Geiger H, Marsden E (1913) The laws of deflexion of α particles through large angles. PM 25: 604

Rutherford E (1914) The structure of the atom. Philosophical Magazine 27: 488-498.

Rutherford E (2009) LXXIX. The scattering of α and β particles by matter and the structure of the atom. The London, Edinburgh, and Dublin Philosophical Magazine and Journal of Science Series 6 21 (125): 669-688. https://doi.org/10.1080/14786440508637080.

Rutherford E (2009) VIII. Uranium radiation and the electrical conduction produced by it. The London, Edinburgh, and Dublin Philosophical Magazine and Journal of Science. Series 5 47 (284): 109-163. https://doi.org/10.1080/14786449908621245.

Sivulka G (2017) An introduction to the evidence for stellar nucleosynthesis. Submitted as coursework for PH241, Stanford University, Winter 2017.

Thomson JJ (2009) XXIV. On the structure of the atom: an investigation of the stability and periods of oscillation of a number of corpuscles arranged at equal intervals around the circumference of a circle; with application of the results to the theory of atomic structure. Series 6 The London, Edinburgh, and Dublin Philosophical Magazine and Journal of Science 7 (39): 237-265. https://doi.org/10.1080/14786440409463107.

Atomic and Nuclear Structure

Key Learning Objectives
- Structure of an atom
- Nuclear structure
- Forces stabilising the atom and nucleus
- Radionuclides

Overview
Atomic structure refers to the structure of an atom that consists of a dense nucleus with electrons revolving around it in specific shells. Nuclear structure means the atomic nucleus is composed of protons and neutrons. All these particles define a nuclide. A set of nuclides with the same number of protons is called an isotope, and if the isotope is unstable, it is called a radionuclide or radioisotope. Each radionuclide emits radiation at its own specific rate and tends to become more stable. There are a number of forces that play a significant role in the stability of the nucleus, and hence the atom.

This chapter briefly reviews some essential aspects of atomic structure for understanding the concepts of radioisotopes.

A. Pathak, *Tools and Techniques in Radiation Biophysics*,
https://doi.org/10.1007/978-981-99-6086-6_2

1 Atom

The matter consists of elements (118 have been identified so far) that, in turn, are made of atoms. Atoms are the smallest known electrically neutral components of an element that retains all the chemical properties of that element. Each atom is composed of a nucleus containing neutrons and protons, and one or more electrons bound to the nucleus and revolving around it in specific shells (Fig. 1).

Neutrons and protons are known as nucleons. Nearly all of an atom's mass is concentrated in its nucleus.

Let us now discuss some of the important properties of these particles.

1.1 Electron

Electrons are subatomic particles with a negative charge of 1.6×10^{-19} coulombs and a mass of 9.108×10^{-28} g. It orbits around the nucleus in different shells. The electron plays a significant role in a number of physical processes such as electricity, magnetism, thermal conductivity, gravitational, electromagnetic, and weak interactions. For example, current is produced by the flow of electrons down wires that powers the electronic devices on which the modern world relies, and this forms the foundation of "microchip" devices which are the type of electronic equipment. A beam of electrons from a heated filament inside a television tube causes the phosphors that are on the front of the tube to glow and illuminate, which is used to create the visible image. Thus, we can say that the electron is the most widely recognised of all the particles because it shows its effects in everyday life.

Fig. 1 Structure of atom

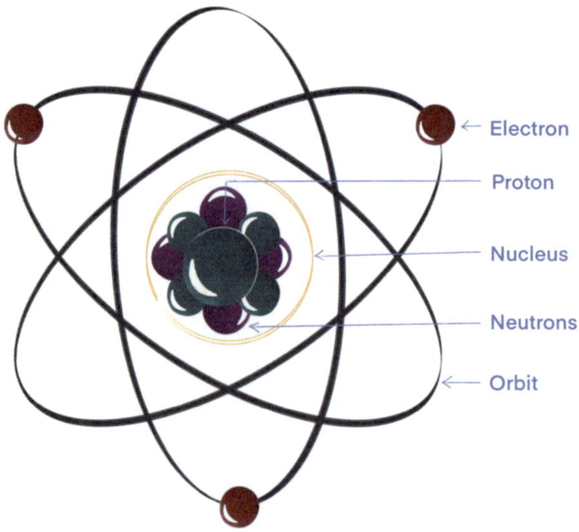

JJ Thomson discovered the electron in a device resembling a television tube called a Crookes tube in 1897. He found that the cathode rays released by a Crookes tube's negative electrode were in fact particles. The size of these particles is so small that it cannot be measured even with the best of modern experimental instruments.

1.2 Proton

The proton is another fundamental particle to be discovered. Protons are particles with the same amount of positive charge as electrons. Wilhelm Wien analysed the rays emitted from a hole in the Crookes tube's negative electrode and discovered the first evidence of its existence in 1898. In 1911, Thomson discovered that the lightest of these positively charged rays. Ernest Rutherford gave the particles the name proton in 1920, when he realised that they were an integral part of all atoms. The proton has a mass of 1.6×10^{-24} g. Therefore, a proton is extremely small, but, unlike an electron, its dimensions can be measured using contemporary technology. The positive charge that a proton possesses is equal to and opposite the negative charge of an electron. The number of protons contained within the nucleus, often known as the atomic number, is the defining characteristic of an element (symbolised by the symbol Z). Because each element has its own number of protons and therefore an atomic number, it specifies the position of that element in the periodic table and, therefore, its chemical identity.

1.3 Neutron

This is another fundamental constituent of an atom, possessing no charge and a mass somewhat greater than that of a proton. In 1932, James Chadwick's experiments led to the discovery of the neutron. The number of neutrons in the atomic nucleus is the neutron number (represented by the symbol N) that determines the isotope or nuclide. The mass number (symbolised by A) equals Z plus N. The neutron possesses a mass of 1.67×10^{-24} g; hence, it is heavier than an electron by a factor of about 2000. Protons and neutrons are bound together within the nucleus with the help of nuclear force, and neutrons are necessary for the stability of nuclei. Neutrons are primarily responsible for the nucleosynthesis of the elements within stars through fission, fusion, and neutron capture processes.

Apart from these three subatomic particles, there are other particles also that are important in radiation biology.

1.4 Positron

Paul Dirac, a theoretical physicist, predicted the presence of positively charged electron-like particles called positrons in 1926. Carl Anderson initially found these positrons in 1932 while examining cosmic rays. Positrons (or anti-electrons or positive electrons) are the antiparticles or anti-matter counters of the electron. They have a mass comparable to that of an electron but have a positive charge. In reality, they are antiparticles of the electron and will be destroyed if allowed to approach an electron, producing gamma rays. Positrons can be produced via positron emission (weak interactions) or pair production when an energetic photon interacts with an atom in a material.

Positrons are essential in positron emission tomography (PET), a diagnostic technology that is particularly useful for imaging the brain and nervous system.

1.5 Neutrino

In 1930, Wolfgang Pauli proposed the neutrino as a possible theoretical explanation for certain radioactive decay observations. In 1934, Enrico Fermi gave the particle the name neutrino. However, it was not demonstrated experimentally until 1956, when nuclear reactors became accessible.

Neutrinos are the elementary particles that constitute the universe. Neutrinos are similar to electrons except that they do not possess an electric charge. Neutrinos are insensitive to the electromagnetic forces that act on electrons since they are electrically neutral. As they are only impacted by a "weak" subatomic force with a far lower range than electromagnetic, they can travel vast distances through matter without being affected by it. Neutrinos have no charge on them and almost no mass, so they interact very weakly with other particles, making them difficult to detect. Neutrinos are produced by dramatic astronomical phenomena such as exploding stars and gamma-ray bursts.

Check Points

- What are nucleons?
- What are positrons used for?
- Why is neutrino so named?

In summary, it can be said that an atom is the basic particle of an element, consisting of a nucleus with positively charged protons and neutrons that are neutral. It is surrounded by a cloud of negatively charged electrons that revolve in discrete shells around the nucleus. Despite being composed of different charges, an atom is electrically neutral. Apart from these, there are other particles, such as neutrinos and positrons, that are produced during atomic and nuclear decay.

1.6 Atomic Structure

After discussing the different components of an atom, let us discuss various features of the atom and its structure. An atom is typically neutral because it has an equal number of electrons and protons. The electrons, protons, and neutrons in an atom are structured like a solar system, with protons and neutrons in the centre and electrons revolving in spherical orbits of varying radii. The nucleus occupies the central position, containing protons and neutrons. The size of atoms in various elements varies widely, ranging from 1 to 2×10^{-8} cm. The nucleus is very small as compared to the atom (i.e., about 10^5 times smaller, or 10^{-13} cm in size).

Key Concept
Modern atomic theory is proposed by John Dalton in 1909.

The electrons revolving in the spherical electron shells are stabilised by Coulombic interactions between the nucleus (positively charged due to protons) and the negatively charged electrons. K-shell is the designation for the shell with the smallest radius, followed by L-shell, M-shell, etc. There is a limit to the number of electrons that can occupy each shell. The K-shell can have up to two electrons, the L shell can accommodate up to eight, the M-shell can have 18, and the N-shell can have up to 32.

Each electron in a particular shell is bound to the nucleus with a fixed amount of energy. Therefore, in order to remove an electron from a particular shell, the electron must receive energy from the outside. The minimum quantity of energy required to remove an electron from a particular shell in an atom is known as the electron's binding energy. The *electron volt* (eV) is the amount of energy gained by an electron when accelerated by 1 volt of potential difference and is used to measure energy on the atomic scale. K-shell electrons are the most tightly bound electrons within an atom, requiring the greatest amount of energy to remove. In contrast, electrons in the outermost shell are the least tightly bound, necessitating the least amount of energy to remove them. Therefore, we can assert that the binding energy of electrons in various shells increases exponentially with Z.

Under normal circumstances, electrons occupy the lowest possible shells (those closest to the nucleus), consistent with the number of electrons that can occupy a given shell. By absorbing energy, electrons can be temporarily shifted to higher shells (unoccupied shells). This absorption can occur in a variety of ways, including by heating a substance, exposing matter to powerful electric fields, passing a charged particle through matter, or even by a strong mechanical impact. When an electron absorbs enough energy to leave an atom, the process is known as ionisation, and the remaining atom is known as an ion. When the electron absorbs just enough energy to move it into a higher, unoccupied shell, the process is called excitation, and the

Table 1 Relative strengths of different types of forces

S. no.	Type of force	Strength
1	Strong	1
2	Electromagnetic	10^{-2}
3	Weak	10^{-13}
4	Gravitational	10^{-39}

atom is considered an excited atom. Excited atoms tend to revert to their normal configurations by emitting electromagnetic radiation (in the form of light, ultraviolet, or X-rays) within 10^{-9} s, on average.

Currently, four types of forces are known: gravitational, weak, electromagnetic, and strong (Table 1).

- *Gravitational forces* are produced as a result of the mass of matter and play a vital role in maintaining the intactness of our solar system, but they are negligible between atoms and molecules.
- *Electromagnetic forces* dominate our daily lives because they hold the atoms together and are responsible for interactions between atoms, molecules, and biomolecules.

Weak forces play a significant role in nuclear transformation. These forces govern the beta decay process and the interactions of neutrinos with nuclei.

- *Strong forces* are the forces that hold a nucleus together and act between proton–proton, proton–neutron, and neutron–neutron.

The interaction strength is proportional to the distance between two protons. Gravitational and electromagnetic forces have an infinite range, and their strengths decrease with increasing separation, r, decreasing as $1/r^2$. However, due to the large mass of the exchange particles for the strong and weak forces, the force associated with them is zero beyond a short distance.

Protons and neutrons are commonly called "nucleons". They are bound in the nucleus by a strong force and have identical, strong, attractive interactions. This powerful force between nucleons has a very short range and is only active at a few femtometers.

Check Points
- Describe the structure of an atom.
- What is the binding energy of the atom?
- Explain the forces that stabilise the atomic structure.

In summary, it can be concluded that the binding energy of an electron in an atom is the least amount of energy required to separate an electron originating from that atom. An atom is stable if it has sufficient binding energy to hold its nucleus together. This is possible when the forces among the various particles in an atom are balanced.

After discussing the atomic structure, the various particles present in the atom, and the different forces stabilising the structure of the atom, let us now concentrate on the nucleus, which will help in building the concept of isotopes.

2 Nuclides

Similar to how various types of atoms are known as "elements", various types of nuclei are known as "nuclides".

A standard symbol for a nuclide is

$$^{A}X_{Z}$$

where A is the mass number, Z is the atomic number (i.e., the number of protons in the nuclide), and X denotes the element to which nuclide belongs.

Key Concept
Nuclide or nuclear species that may or may not be the isotopes of a single element.

An *element* is defined solely by its atomic number (Z), whereas a nuclide has a specific *mass number* (A) and an *atomic number* (Z). Atoms are electrically neutral because they contain an equal number of protons and electrons. The sum of protons and neutrons is known as the mass number (already discussed). Since the number of protons is unique to each element, each nuclide is specified by its element and its mass number. The nuclide's mass number (A) equals the sum of its protons (Z) and neutrons (N), i.e.,

$$A = Z + N \tag{1}$$

For example, $^{131}I_{53}$ a nuclide of iodine, consists of 53 protons and 78 neutrons, and 131 is the mass number.

Protons and neutrons have nearly identical rest masses, and the proton possesses a positive charge of the same magnitude as the electron's negative charge, whereas the neutron has no charge.

The majority of the mass of the atom is concentrated within the nucleus, which is composed of a Z number of protons and an $A–Z$ number of neutrons.

On the basis of their mass number, neutron number, and atomic number, nuclides are classified as follows (Table 2).

Table 2 Properties and examples of three different types of nuclides

S no	Nuclide	Property	Example
1	Isotopes	Same number of protons	$^{123}I_{53}$, $^{124}I_{53}$, $^{125}I_{53}$, $^{126}I_{53}$, $^{127}I_{53}$, $^{128}I_{53}$, $^{129}I_{53}$, $^{130}I_{53}$
2	Isobars	Same atomic mass	$^{131}Sn_{50}$, $^{131}Sb_{51}$, $^{131}Te_{52}$, $^{131}I_{53}$, $^{131}Xe_{54}$, $^{131}Cs_{55}$
3	Isotones	Same number of neutrons	$^{125}Cd_{48}$, $^{126}In_{49}$, $^{127}Sn_{50}$, $^{128}Sb_{51}$, $^{129}Te_{52}$

2.1 *Isotope*

Isotopes are two or more species of atoms of the same chemical element that share the same atomic number, the same location in the periodic table, and nearly identical chemical properties but distinct atomic masses and physical properties. The name "isotope" derives from the Greek words iso (means same) and topos (means place) and thus suggests that different isotopes of an element occupy the same position in the periodic table. In other words, an element can have atoms with the same number of protons (or atomic number Z) but different numbers of neutrons (or mass numbers A). $^{13}C_6$ and $^{14}C_6$ are examples of carbon isotopes. When the number of neutrons varies, so will the mass number. In 1913, Soddy was the first to demonstrate the existence of atoms of the same element with different atomic weights, which he termed isotopes. There may be several isotopes of an element. For example, calcium contains six isotopes; carbon has three; and sodium has only one isotope. An isotope is typically represented by the symbol of its element with a superscript of its mass number. For example, ^1H (common hydrogen), ^2H (deuterium), and ^3H (tritium) are isotopes of hydrogen; ^{10}C–^{16}C are isotopes of carbon; ^{32}P is an isotope of phosphorus; and ^{35}S is an isotope of sulphur.

2.2 *Isotones*

Isotones are nuclides containing an identical number of neutrons. For example, $^{13}N_7$ and $^{14}O_8$—both of these nuclides contain six neutrons. ^3H (tritium) and ^4He are examples of isotones, each having two neutrons ($A - Z = 2$).

2.3 *Isobars*

Isobars are the isotopes of various elements with the same mass number. For example, cobalt-60 and nickel-60 are isobars each with 60 nucleons ($A = 60$). $^{131}I_{53}$ and $^{131}Xe_{54}$ are other examples of isobars.

Like an atom, the nucleus also exists in an excited state for some period of time, and therefore, it is said to be in an isomeric (metastable) state.

Check Points
- What are nuclides?
- What is the difference between a nuclide and an isotope?
- Define isotones and isobars.

In summary, we can say that nuclides are species having a specific number of nucleons. If the number of neutrons varies, they are called isotopes. These isotopes may be stable or unstable. If they exist in an unstable form, they become radioisotopes or radionuclides that have the property of continuously emitting energy in the form of radiation.

3 Nuclear Structure

We have already discussed that within an atom, electrons can only exist in well-defined energy shells. Similarly, the nucleons within a nucleus exist at distinct energy levels. However, the situation inside the nucleus is slightly complicated due to the presence of two types of nucleons—protons and neutrons—that must coexist. Similar to electrons, nucleons can be excited to higher shells by absorbing energy from the outside. Neither the mechanism by which nucleons occur in these shells nor the transition between these shells are well understood. The lowest possible configuration of nucleons within the nucleus of a nuclide is called the ground state. Excited state is the common name for the higher shells. Similar to the electrons in an atom, nucleons within the nucleus are also bound with different binding energies. Unlike electrons, which are bound by electrostatic forces, nucleons are stabilised by more powerful nuclear forces. These attractive forces are sufficient to overcome the electrostatic repulsions that exist between positively charged protons. Similar to the excited states of an atom, which decay to the ground state by releasing light energy or X-rays, the excited states of nuclides also decay to the ground state by the emission of high-energy radiation. Thus, nuclides and isotopes are identical in every way except the amount of nuclear energy they contain. In general, the nuclides in the excited states are of very short duration (less than 10^{-11} s). In a few instances, though, nuclides stay in the excited state for a very long time (seconds, minutes, or even longer, i.e., years). Under such conditions, the excited state is called a metastable state. For example, ^{99m}TC, ^{113m}In. The letter "m" after the mass number denotes the metastable state. Because the excited states of the nuclides have the same mass number (*A*), same atomic number (*Z*), and the same number of neutrons as the ground state, these nuclides are called isomers (i.e., an isomer is the nuclide's excited state). Examples include ^{99m}Tc (an isomer of technetium-99) and ^{60m}Co (an isomer of cobalt-60).

Key Concept
Ernest Rutherford postulated the nuclear structure of the atom.

The binding energy of a nucleon (i.e., the quantity of energy required to extract it from the nucleus) varies among nuclides. The sum of the masses of the nucleus composed of Z protons and $(A-Z)$ neutrons is greater than the mass of the nucleus itself. This difference in masses is known as the mass defect (deficit) Δm and its energy equivalent Δmc^2 is called the total binding energy (EB) of the nucleus. Consequently, the *total binding energy* (EB) associated with a nucleus may be defined as:

- The amount of positive work necessary to breakdown a nucleus into its constituent protons and neutrons.
- The amount of energy released when Z protons and $(A–Z)$ neutrons are combined to form the nucleus.

The *binding energy per nucleon* (EB/A) in a nucleus (i.e., the total binding energy of a nucleus divided by the number of nucleons) fluctuates with the number of nucleons A and is of the order of 5–8 MeV/nucleon. This is about a thousand times more than the average binding energies of electrons in atoms. In order to remove a proton or neutron from a nucleus, a substantial amount of energy must be transferred from outside, which is generally only achievable in nuclear reactors, accelerators, or cyclotrons.

Key Concept
Nuclides containing an even number of both protons and neutrons are most stable, i.e., less radioactive.

4 Radionuclides

Although many nuclides are stable in their ground state, most are unstable and are called radionuclides. Radionuclides (or radioisotopes) are isotopes that are unstable, or radioactive, and give off radiation spontaneously. Approximately 254 stable isotopes and 50 radioisotopes are naturally occurring (i.e., *radioactive*). Thousands of other radioisotopes have also been produced in the laboratory. Nuclear reactors produce numerous radioisotopes by bombarding suitable objects with readily available neutrons. However, some are also produced by the interaction of protons, deuterons, or other subatomic particles that have been accelerated to high speeds using a cyclotron or similar accelerator.

The equilibrium between the number of neutrons and the protons within a nucleus determines the stability of the radioisotopes. There are two types of forces that determine the overall stability of a nuclide:

4.1 Strong Forces

As previously mentioned, nucleons are held together by strong forces (e.g., between proton–proton, proton–neutron, and neutron–neutron). These are the attractive forces that exist only when the distance between two nucleons is extremely small.

4.2 Electromagnetic (em) Forces

These forces only work between protons because neutrons don't have any charge. These forces push against each other because similar charges repel each other. The equilibrium between these two forces determines the stability of a nuclide. Whenever the equilibrium between these two opposing forces is disturbed, the nuclide becomes unstable and therefore radioactive.

Radionuclides attempt to establish stability by releasing electromagnetic radiation or charged particles. This release is called radioactive decay. As previously discussed, the stability of an isotope of a particular element is also related to the ratio of the number of neutrons to the number of protons in the nucleus; this is the primary cause of radioactivity. This relationship between the number of protons and neutrons in the nucleus can be observed by plotting the values of N versus Z for known isotopes (Fig. 2). This curve starts with a straight line and then gradually curves towards the neutron number as the atomic number increases.

The total number of protons and the repulsive force acting on these protons increase as the atomic number increases. Nevertheless, because the nuclear force has a short range, there is a point at which the electric force (which does not have a short range) becomes so great that it begins to balance the nuclear force for a proton. This condition is precarious. Since there is no increase in electric force, this condition would not arise if the number of neutrons within a nucleus were to be increased.

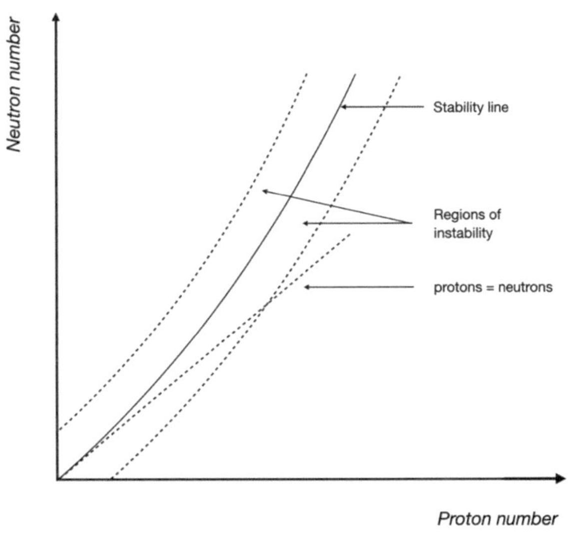

Fig. 2 Relationship between number of protons and neutrons

Therefore, when $N > Z$, larger nuclei are more stable. For nuclei with a Z value greater than 83, however, the addition of neutrons will not produce a stable isotope because the electric force is too strong and overcomes any additional (glue) force acting between neutrons and protons. These nuclei are unstable.

Figure 2 shows the graph with the number of neutron (N) against the vertical axis and the number of protons (Z) against the horizontal axis. The solid black line represents the non-radioactive nuclides that lie along a stability diagonal. For lighter nuclides (i.e., $A = 50$), the number of protons and neutrons is equal in stable nuclides; for instance, the stable oxygen (O) nuclide ^{16}O contains 8 neutrons and 8 protons. Stable nuclei have roughly equal numbers of protons and neutrons near the bottom of the graph, but as the graph rises, stable nuclei require an increasing number of neutrons relative to protons. Nuclei with an even number of protons and neutrons are more likely to be stable than those with an odd number of protons and neutrons. Stable nuclides contain the optimal number of protons and neutrons, which reduces nuclear energy. The lines that remain represent radionuclides or radioactive nuclides. These are unstable because their combination of protons and neutrons is not optimal. They will decay via one of the several distinct mechanisms, transforming into a number of nuclides until the equilibrium line is reached. However, for heavier nuclides (i.e., with $A > 100$), significantly more neutrons than protons are necessary for the stability of a nuclide. i.e., the stability of elements with higher atomic numbers is frequently accompanied by a neutron-to-proton ratio greater than one. If a radionuclide is located in the upper region of the curve, it has an excess number of neutrons that contribute to its instability and decay by positron emission (also known as *beta plus decay*) or *electron capture*. Alternatively, if the radionuclide is located in the lower region of the curve, the excess of protons makes the nuclide unstable and disintegrate through electron emission (also called *beta-minus decay).* It may be inferred that the N:Z ratio of a stable isotope lies within restricted limits; an isotope outside these ranges will be unstable. Such a nucleus will emit radiation as it attempts to alter its N:Z ratio towards stability. Therefore, the phenomenon of radioactivity might be understood as the isotope's endeavour to achieve stability. These decay processes will be discussed in subsequent chapters. The chart of the nuclides can be compared to the contour map of a steep-sided valley, which depicts the route by which boulders will find their way down the slope, losing energy until they reach the valley floor.

Check Points

- How is the binding energy of a nucleus related to its stability?
- What are radionuclides?
- Explain the forces stabilising the nuclide.
- How is the stability of an isotope related to the number of protons and neutrons in the nucleus?
- Explain the E Rutherford's experiments to invent different types of radiations.

In summary, it can be said that the ratio of neutron to proton and the total number of nucleons present in the nucleus are the two main factors that affect the stability of the nucleus. The lighter elements ($Z < 20$) have a ratio of 1:1 and are more stable.

Further Reading

Alford TL, Feldman LC, Meyer JW (2007) Fundamentals of Nanoscale Film Analysis. Chapter: An Overview: Concepts, Units, and the Bohr Atom pp 1–11. Springer US. ISBN: 978-0-387-29261-8

Bauer R, Birk J, Marks P (2019) Introduction to Chemistry. 5th edn. McGraw-Hill Education. ISBN 978-1-259-91114-9

Demtröder W (2022) Nuclear and Particle Physics. Springer Cham. ISBN: 978-3-030-58313-2. doi:https://doi.org/10.1007/978-3-030-58313-2

Introduction to Radiation (2012) Published by the Canadian Nuclear Safety Commission (CNSC). ISBN 978-1-100-21572-3

L'Annunziata MF (2020) Chapter 1 - The atomic nucleus, nuclear radiation, and the interaction of radiation with matter. Handbook of Radioactivity Analysis (4th edn.) Volume 1: Radiation Physics and Detectors. Pages 1-243. eBook ISBN: 9780128143988. Hardcover ISBN: 9780128143971. doi:https://doi.org/10.1016/B978-0-12-814397-1.00001-7. Science Direct. Academic Press

Islam MR, Hossain ME, Islam AO (2018) Hydrocarbons in Basement Formations. Scrivener Publishing LLC. Print ISBN:9781119294221. Online ISBN:9781119294498. DOI:https://doi.org/10.1002/9781119294498

Keighley J (1990) Work Out Physics GCSE. Red Globe Press London. doi:https://doi.org/10.1007/978-1-349-11839-7

Lawson R (1999) An introduction to radioactivity. Nuclear Medicine Department, Manchester Royal Infirmary KIPDF pp1-20

Levin F (2006) Calibrating the cosmos: How cosmology explains our big bang universe. Part of the Astronomers' Universe Series book series (ASTRONOM). pp 37–59. Chapter: Light, Radiation, and Quanta. Springer. ISBN: 978-0-387-49768-6

Miller MA, Carey A (2001) Isotopes, Natural. doi:https://doi.org/10.1002/14356007.a15_001. Wiley Online Library

Podgorsak EB (2010) Radiation Physics For Medical Physicists. Biological and Medical Physics, Biomedical Engineering 2nd edn. Springer-Verlag Berlin Heidelberg. ISSN 1618-7210. ISBN 978- 3-642-00874-0. e-ISBN 978-3-642-008745-7. DOI https://doi.org/10.1007/978-3-642-008745-7

Podgorsak EB (2012) Chapter 1: Basic Radiation Physics. Review of Radiation Oncology Physics: A Handbook for Teachers and Students. IAEA publication. ISBN 92-0-107304-6

Podgorsak EB (2014) Compendium to Radiation Physics for Medical Physicists. 300 Problems and Solutions. Springer Berlin, Heidelberg. ISBN: 978-3-642-20186-8. doi:https://doi.org/10.1007/978-3-642-20186-8

Podgorsak EB (2016) Radiation Physics for Medical Physicists. 3rd edn. Springer. ISBN: 978-3-319-25382-4

Semat H, Albright JR (1972) Introduction to Atomic and Nuclear Physics. 5th edn. Springer. ISBN: 978-1-4615-9701-8

Wackers FJT (ed) (1980) Thallium-201 and Technetium-99m-Pyrophospate myocardial imaging in the coronary care unit. Part of the book series: Developments in Cardiovascular Medicine (DICM, volume 9). Springer. ISBN: 978-94-009-8904-7

Waight N, Liu X, Gregorius RM, Smith E, Park M (2013) Teacher conceptions and approaches associated with an immersive instructional implementation of computer-based models and assessment in a secondary chemistry classroom. International Journal of Science Education 36, 2014 (3): 467-505. doi:https://doi.org/10.1080/09500693.2013.787506

Weiner ER (2008) Applications of Environmental Aquatic Chemistry A Practical Guide. 2nd edn. CRC Press Taylor & Francis Group

Zabel H (2023) 5 Nuclei and isotopes. Volume 2 Physical Aspects of Diagnostics. Published by De Gruyter. doi:https://doi.org/10.1515/9783110757095-005

Radioactivity and Its Units

Key Learning Objectives

- Radioactivity and its causes
- Types of radiations
- Ionising and non-ionising radiations
- Quantity and units of radiations

Overview

Radioactivity is the release of energy due to the nuclear decay of specific types of atoms and isotopes. It is a physical phenomenon that can be determined by taking a count of the number of atoms that spontaneously decay each second. A variety of units are used to measure the radiation dose and exposure. There are different types of radiation of which ionising radiation is the most damaging as it causes numerous health hazards. During the decay of radioisotopes, harmful alpha particles, beta rays, and gamma rays are emitted.

Radioactivity is the property of spontaneous disintegration possessed by certain unstable nuclei (radionuclides), and this disintegration is always followed by the release of high-energy radiation that includes alpha or beta particles or gamma rays. Radioactivity may also be defined as a process that involves the disintegration of an unstable parent nucleus disintegrates into a potentially stable daughter nucleus. This unstable daughter nucleus will continue to decay until a stable configuration is attained. This process remains uninfluenced by temperature, pressure, chemical combinations, etc., that are involved in the regulation of the rates of chemical reactions. The rate at which radioactive decay occurs has been found to be influenced by

A. Pathak, *Tools and Techniques in Radiation Biophysics*,
https://doi.org/10.1007/978-981-99-6086-6_3

the structure of the unstable nucleus. These radiation can produce damage to living cells, but they are also involved in a vast array of positive applications, particularly in the area of medicine, when handled appropriately. The released energy can be harnessed as nuclear power, or if released by radioactive contamination in the environment, it can prove to be dangerous beyond imaginable magnitudes.

This phenomenon of spontaneous disintegration occurs naturally in a number of substances. For instance, since the formation of the earth, radioactivity has been found in many natural materials on it (like ^{40}K forms part of our bodies). However, these radiation cannot be detected normally. The phenomenon of radioactivity was discovered only 100 years ago after the development of radiation detectors. Today, techniques have been developed to create new radioactivity sources; some of these (for example, ^{131}I and ^{99}Mo) are unintentional byproducts in the nuclear power sector that have several important medical applications, while other radioisotopes (for example, ^{18}F) are produced specifically for the benefits of their medical use. All elements ranging from atomic number 83 (bismuth) to 92 (uranium) are radioactive because the elements with larger nuclei often tend to be unstable, as explained below. There are other elements that are also radioactive but are outside the range of 83–92 protons.

1 Causes of Radioactivity

Key Concept

The average annual natural radiation exposure reported for India is 0.73 mSv. Out of this, 0.355 mSv per year comes from cosmic rays, and rest is 0.379 mSv per year from terrestrial components.

Since radioactivity is due to the disintegration of atomic nuclei, the factors that destabilise the nucleus enhance the radioactivity. The ratio of the number of the neutrons to the number of protons (N/P ratio) is one of the major factors in determining disintegrations. For nuclei with an atomic number of 20 or less, an N/P ratio indicates stability. If the number of protons is greater than the number of neutrons, or vice versa, the nucleus disintegrates by ejecting the additional component as radiation and thus achieves stability. For nuclei with an atomic number above 20, stability is ensured only when the neutron number slightly exceeds that of the protons. It is believed that when nucleus contains a large number of protons, they repel each other and thus lower its stability. The forces of repulsion can be diminished and the stability enhanced to some extent if the number of neutrons is increased. This causes an increase in the size of the nucleus, and hence the protons tend to lie further apart. However, the nucleus becomes unstable when there is a large rise in the proton or neutron number. The nucleus then undergoes radioactive disintegrations, giving out α- or β-particles in such a manner as to yield a stable nuclear reactor or accelerator.

2 Categories of Radiation

As discussed earlier, radioactivity and, therefore, radiation are natural components of the earth's natural environment. Radiation is part and parcel of life, i.e., our planet is innately radioactive. Cosmic rays and some natural materials also cause unavoidable exposure to radiation. Therefore, radiation are grouped into natural and man-made. Humans are exposed to radiation that comes from several sources that are discussed below.

2.1 Natural Radiation

Key Concept
Natural background radiation contribute around 81% of the annual dose of the population and man-made background radiation contributes the remaining 19%.

Natural or background radiation are those radiation that are constantly found in the atmosphere, and are released from several sources. According to the United Nations Scientific Committee on the Effects of Atomic Radiation (UNSCEAR), there are two primary sources of natural radiation:

2.1.1 Cosmic Radiation

Cosmic radiation are extremely high-energy radiation that originate in space, beyond the earth's atmosphere, and are primarily composed of charged particles. This includes the radiation coming from the sun and the secondary radiation produced by the entry of charged particles with high velocities and energies into the earth's atmosphere. Cosmic radiation releases radioactive isotopes like ^{3}H (tritium), ^{7}Be, ^{14}C, and ^{22}Na into the atmosphere. The acute and chronic health effects of these radiation involve both direct damage to the genetic material and indirect effects produced due to the formation of reactive oxygen species. Cosmic radiation is also found to cause the ionisation of nitrogen and oxygen molecules in the atmosphere, which ultimately leads to the chemical reactions that may result in ozone depletion.

2.1.2 Terrestrial Radiation

This type of radiation is produced by radioactive substances that were present in the earth's crust five billion years ago. Radioactive elements that are commonly found on earth and contribute to terrestrial exposure are isotopes of potassium, thorium, actinium, radon, and uranium and their decay products. Most of these sources have been decreasing due to radioactive decay that has been taking place since the formation of the Earth. The level of radioactivity over the water bodies tends to be about a tenth of the background terrestrial radiation. However, the coastal areas (including areas next to fresh water) may get an additional contribution from the distributed sediments. There is a minimal effect of these isotopes on humans because of the actual diminishment of these isotopes due to their decay.

2.2 *Man-Made Radiation*

Man-made or technogenic radiation are those that are produced by humans, whether for their own medical use, from nuclear tests, or from television and other devices like smoke detectors.

2.2.1 Internal Radiation

Swallowed, breathed, absorbed radionuclides leads to internal exposure. The majority of the energy of the radiation will be absorbed by cells, tissues, and organs once it enters the body. Terrestrial radionuclides absorbed by plant roots may be integrated into food. Radionuclides can be ingested by humans when contaminated plant material or animals that eat contaminated plant material is consumed. Inhalation of radioactive decay products of naturally occurring radon gas is another method of exposure. The level of an internal radiation dose is proportional to the amount of emitted radiation and the type of emitted radiation. Internal radioactive substances emit radiation until they cease to be radioactive (i.e., they decay) or until they are naturally eliminated by the body.

2.2.2 Medical Radiation

Radiation is routinely used in medical procedures, including chest as well as dental X-rays, CT scans, mammography tests, thallium stress tests for heart, tumour irradiation therapies, and other diagnostic imaging. Radiopharmaceuticals are used to support these diagnostic tests. Physicians also use radiation to carry out certain procedures, like heart valve replacement. However, the benefits from medical procedures vastly outweigh the small risk of harm from the radiation doses employed. This is the reason that these procedures are still being used in the medical field.

2.2.3 Anthropogenic Radiation

Consumer products like static eliminators (which contain the isotope ^{210}Po), smoke detectors (which contain the isotope ^{241}Am), cardiac pacemakers (which contain the isotope ^{238}Pu), fertilisers (which contain uranium and thorium decay series isotopes), and tobacco products (which contain the isotopes ^{210}Po and ^{210}Pb), are all sources of anthropogenic radiation. This category also includes radioactive wastes generated by nuclear power plants, activities like uranium mining and milling, low exposure from the transportation of radioactive materials, scientific research facilities, and radioactive fallout from nuclear weapons testing. Contaminated food and drink is a significant consequence that can occur after a nuclear disaster, even at large distances from the accident site and across international borders. This can affect the lives of a large number of dependent communities. According to the World Health Organisation, it has been found that natural radiation exposure varies depending on geography, social habits (for example, frequent flying), health, occupation, and proximity to routine and accidental releases (Table 1). Other additional sources of anthropogenic radiation include materials used in construction of buildings and roads, combustible fuels, X-ray security systems, televisions, luminous watches, smoke detectors, cigarette products, ophthalmic glass used in eyeglasses, and some ceramics. This additional exposure to humans above naturally produced background radiation is a reason for concern since it may cause diseases like cancer, etc. Radiation exposure from the average annual dose from all natural sources is roughly 2.4 mSv, although this figure can vary significantly based on geographical location.

Check Points

- Define radioactivity.
- What are the causes of radioactivity?
- State the sources of natural radiation.
- Where do man-made radioactive substances come from?

It can be summarised that there is spontaneous disintegration of isotopes followed by the emission of high-energy radiation into a more stable state. This

Table 1 Approximate relative contributions of radiation exposure from different sources

S. No.	Source	Percent total dose
1	Radon (natural internal exposure)	43
2	Earth gamma radiation (Natural internal exposure)	15
3	Artificial sources (Medical exposure, X-rays, etc.)	20
4	Cosmic rays	13
5	Food/water	8
6	Others (all man-made sources)	1
	Total	~100

disintegration is affected mainly by the ratio between the number of neutrons and protons within the nucleus.

Thus cosmic rays and radionuclides in the earth's crust, air, food, and water, as well as the human body itself, are the primary contributors to human radiation exposure. Exposure to man-made radiation occurs primarily through the use of radiation and radioisotopes in medicine, as well as through occupational sources in the production of energy from nuclear power reactors, nuclear processes in industry, and even through nuclear weapons testing in the past. However, it has been found that most of these man-made exposures do not possess any health hazards, are of very low intensity, and also have a low total dose.

3 Classification of Radiation Based on Different Properties

As already known, radiation deals with the emission of energy either through space or a medium as waves or particles. Radiation is characterised by its type and the energy associated with it (Fig. 1). There are two main categories of radiation: particulate and electromagnetic.

3.1 Particulate Radiation

These radiation are made up of particles with mass and energy that may or may not be electrically charged. Examples of particulate radiation include alpha rays, protons, beta rays, and neutrons.

3.2 Electromagnetic Radiation

These consist of photons that possess energy but have neither mass nor charge. A photon is a "particle" or "quantum" that contains a discrete amount of electromagnetic energy and travels at the speed of light, i.e., 3×10^8 m/s. A photon is also

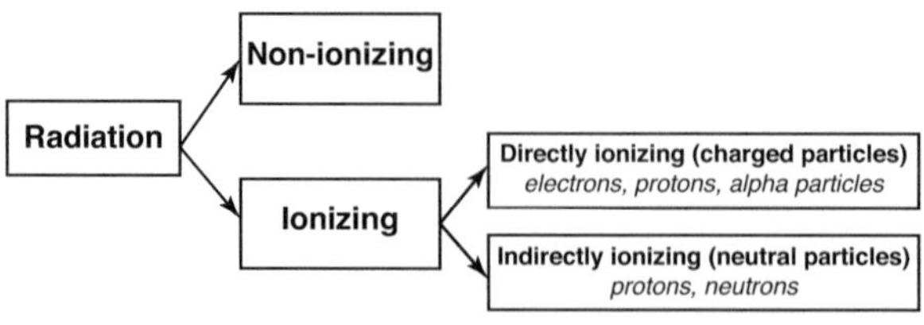

Fig. 1 Classification of radiation. Adapted from International Atomic Energy Agency, Podogorsak, E.B., "Basic radiation physics", Radiation Oncology Physics: A Handbook for Teachers and Students, IAEA, Vienna (2005) 1–43

described as a "packet of light". Examples of electromagnetic radiation are X-rays and gamma rays. visible light, ultraviolet light, and radio waves.

Depending on the energy associated and the capacity to penetrate and ionise the matter, radiations are also classified as ionising radiation and non-ionising radiation. Different types of radiation have different properties.

3.3 *Non-ionising Radiation*

These types of radiation cannot ionise matter because their kinetic energy is less than the ionisation potential of matter. The ionisation potential is the minimum amount of energy necessary to ionise an atom, and its value can be from a few eV for alkali metals to 24.6 eV for the noble gas helium). These radiation can only shake or move molecules, which can mainly lead to a temperature increase. Therefore, this type of radiation is harmful to cells only if its intensity is sufficient that it causes excessive heating. Non-ionising radiation is not harmful. Microwaves, radio waves, infrared rays, ultraviolet light, and visible light—all these fall into the category of non-ionising radiation.

3.4 *Ionising Radiation*

These radiation have sufficiently high energy (more than 10 eV) to cause ionisation of the atoms and molecules and break chemical bonds, thus causing unpredictable chemical reactions. This type of radiation causes ionisation of the matter either directly or indirectly as its energy level exceeds the ionisation potential of the matter and therefore causes more damage to the biological system as compared to particulate radiation or electromagnetic radiation. These radiations affect living cells, especially the DNA, thus increasing the risk of cancer. Most ionising radiation originates from radioactive materials. Ionising radiation includes both energy waves and particles. Therefore, there are two major categories of ionising radiation:

3.4.1 Directly Ionising Radiation (Charged Particles)

Some of the examples of these radiation include electrons, protons, heavy ions, and alpha particles. Here, the particles are moving at a rapid rate. Ionising radiation deposits energy via the charged particle interaction with the orbital electrons of atoms in the medium that it is passing through coulomb forces. Consequently, the vast majority of particulate radiation is directly ionising that individual particles with enough kinetic energy can directly alter the atomic structure of the absorbing media they pass through, damaging molecules chemically and biologically.

3.4.2 Indirectly Ionising Radiations (or Neutral Particles)

These include photons (that includes X-rays, gamma rays, and UV rays) and neutrons. Indirectly, ionising radiation interacts with the medium in two steps. The first stage involves the release of the charged particle into the medium (i.e., electrons or positrons are released by photons, and protons or heavier ions are released by neutrons). In the second stage, energy is deposited in the medium by direct Coulomb interactions between the discharged charged particles and the orbital electrons of the atoms in the medium. These are called indirect ionising radiations, as they themselves do not cause any chemical or biological damage but release secondary electrons (or charged particles) after absorbing energy in the material.

Key Concept
Since directly ionising radiation are charged, they cause ionisations to occur at frequent intervals, as opposed to indirectly ionising radiation that can cause damage by penetrating deeper into tissue and interacting with internally produced directly charged particles.

UV radiation possesses characteristics of both forms of radiation. Despite the fact that the non-ionising component of the UV spectrum enters the atmosphere of the earth, more molecules in biological systems are harmed by radiation than can be explained by thermal effects, such as sunburn, which are common. These properties are due to the fact that UV radiation causes alteration in chemical bonds, even though these radiations do not have enough energy to ionise the atoms.

Therefore, the use of ionising radiation is growing all the time because it has so many useful new uses in health care. These are useful in several branches of medicine, i.e., diagnosis, treatment, and research. The branch of medicine that is concerned with the use ionising radiation for the treatment of disease is called radiotherapy, therapeutic radiology, or radiation oncology, where radiation is specially used for cancer treatment. During this process, radiation causes ionisation of the cells and tissues as it passes through them, which can lead to cell death or changes in the genes, thereby causing mutations. Diagnostic radiology and nuclear medicine are medical specialities that employ ionising radiation for the diagnosis of disease. Doctors can get useful information related to the diagnosis of various diseases by injecting a radioactive substance and monitoring the radiation given off as it travels through the body. These radiation exposures can be markedly reduced by taking proper safety precautions and optimising radioisotopes-based procedures and practices.

Apart from the above-mentioned benefits, these ionising radiations are also responsible for numerous health hazards if not used properly and if proper precautions are not taken. Thus, all types of radiation have numerous beneficial and harmful effects, but we will be concentrating on ionising radiation, mainly the radiation emanating from radioactive materials.

4 Classification on the Basis of their Behaviour in the Electric Field

The rays from a radioactive material can be divided into three types based on their behaviour when placed under the influence of an electric field (Fig. 2).

- **Alpha rays**: These are positively charged and are deflected towards the negative plate.
- **Beta rays**: These are deflected towards the positive plate due to their negative charge.
- **Gamma rays:** These are neutral and therefore remain undeflected in an electric field.

4.1 *Alpha (α) Rays*

> **Key Concept**
> Alpha particles were the first nuclear radiation to be discovered.

These rays consist of particles carrying two protons and two neutrons. In fact, they are the helium (He) nuclei, having a binding energy of 7 MeV per nucleon and a stable configuration. These particles have a velocity of over 16,000 km/s, ionise the

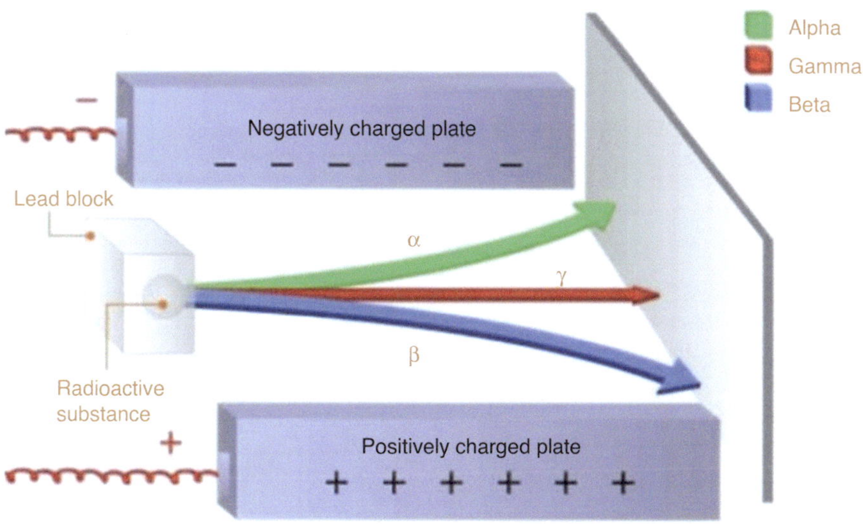

Fig. 2 Different types of radiation travelling through an electric field

air, and cause luminescence on a zinc sulphide (ZnS) screen. A single alpha particle of 5 MeV produces about 25,000 ion pairs per cm of travel. The penetrating power of these particles is less, i.e., they cannot travel very far, and even a sheet of paper, skin, or aluminium foil of 0.1 mm thickness can stop these particles (Fig. 3). They have a range of 3.5 cm in the air. Because of the limited range of alpha particles in matter, they pose no external radiation threat to humans and do not cause any damage to living tissues, as these particles released from alpha decay are not able to penetrate even the outer layers of dead skin cells. However, once they enter the body through inhalation or ingestion, they are very harmful because these powerful particles can do significant harm to the cells they penetrate. They can even rip through cells in organs or blood, releasing their energy and causing extensive injury to the surrounding tissue. Alpha particles are many times more effective at causing cell damage as compared to X-rays and gamma rays. This means a single alpha particle can deliver a substantial dose of radiation to a cell. Alpha emitters are thus an internal hazard, and their ingestion must be prevented. Alpha rays are also released by naturally occurring radioactive substances such as uranium and radon. The radionuclides of heavy elements such as plutonium, polonium, radium, and americium are also involved in the emission of alpha particles.

Fig. 3 Comparison of the penetrating ability of the three types of radiation

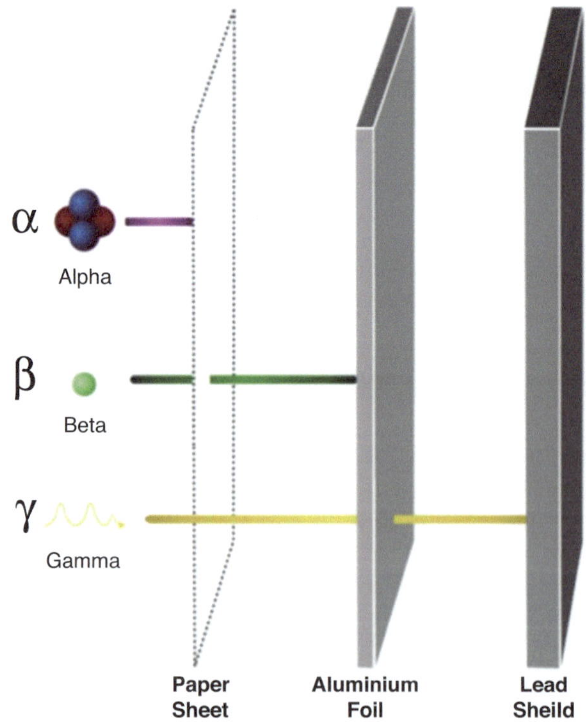

4.2 Beta (β) Particles

Beta particles are basically electrons carrying a single negative charge. Compared with α-particles, these particles (or electrons) are very small, i.e., they are only a fraction of the size of the alpha particles, and therefore they move very fast, having a velocity of 160,000–240,000 km/s. They also ionise air, but owing to their high velocity and small size, they have less possibility than α - particles to interact with matter, and so these are less ionising and thus have a lesser power of luminescence on a ZnS plate. A 0.5 MeV β-particle has a range of 1 m in air and produces about 60 ion pairs per cm of travel. The kinetic energy of β-particles is much lower than α-rays. The maximum energy recorded for a β-particle is 0.54 MeV. Being lighter and faster, β-rays have a greater penetrating power than α-rays. These radiations are largely absorbed by an aluminium foil of 5 mm thickness, glass, or ordinary clothes (Fig. 3). Beta radiation can reach the germinal layer of the human skin. If β emitters are left on the skin for a long time, they may induce skin damage. Due to its low specific ionisation, a beta particle has a larger range in tissue than an alpha particle; hence, it releases less energy per unit volume of tissue and causes less injury than an alpha particle can. The beta particles can also pose an internal threat. External exposure can result in the penetration of beta rays through the surface of the skin to the most sensitive layers. The greatest risk is posed by beta- emitting radionuclides that are inhaled or ingested. Thus, depending on their energy level, beta rays are dangerous both outside and inside the body. ^{90}Sr and ^{3}H (tritium) are examples of beta-emitting radionuclides that are produced by nuclear power plants. Every one of us is exposed to ^{90}Sr as a result of nuclear bomb testing. Our bodies frequently misidentify ^{90}Sr as calcium, accumulating it in the bones that produce new blood cells. Once there, it increases the risk of malignancies of bone and blood, such as leukaemia. Tritium is radioactive hydrogen that binds in a similar manner to normal hydrogen. Hydrogen is the most plentiful of all the elements and also a component of water, which protects our genetic material. Tritium can bind with water, thus irradiating DNA from a very close range. Another example is ^{40}K, which is a natural radioactive element emitting beta radiation.

4.3 Gamma (γ) Rays

These rays are high-energy photons and have no electric charge or mass. Gamma rays are produced due to the atomic disintegrations that take place after the emission of α or β particles. The emission of α and β particles leaves the atomic nuclei in an excited state. The excess energy present in the atom or nucleus is released in the form of γ radiation. Therefore, gamma radiation is released during the nuclear process to get rid of the excess energy associated with the unstable nucleus. They have similarities with X-rays as both can pass through living materials easily, but they have much shorter wavelengths (10^{-10} to 10^{-13} m). They rarely collide with neighbouring atoms and travel vast distances before dissipating all their energy. So γ-rays have an extremely high penetration ability. They can even pass through a steel plate of 0.15 m thickness, and only thick lead blocking their path can stop these radiations (Fig. 3). Dense substances like lead and concrete are effective γ-rays blocking materials. The energy content varies from 0.012 to 1.24 MeV. Gamma rays have enough energy to ionise materials and are therefore capable of damaging live cells. The damage produced by the gamma rays is directly proportional to the number of ionising paths generated in the absorbing material. The harmful effects of these rays depend on their quantity, their energy, and their distance from their source. The intensity of radiation decreases exponentially with an increase in distance from the source.

Key Concept
Alpha particles are the most harmful internal hazard i.e., when swallowed, inhaled, absorbed, or injected while gamma rays are the most harmful external hazards.

^{137}Cs is an example of a gamma emitter that is produced in nuclear reactors. It is similar to potassium and is therefore collected in muscle. Other gamma emitters produced in nuclear plants are ^{131}I and ^{129}I which are used in bomb testing. Radioactive iodine concentrates in the thyroid and emits both beta and gamma radiation in the surrounding tissue.

Gamma rays are harmful because they come from the outside and can easily get into living things and do their damage without being taken in. However, alpha particles and beta rays belong to the category of internal emitters, and they produce the greatest harm to organisms when taken internally (Table 2).

Thus, if different types of radiation are compared, alpha particles (helium nuclei) are found to have a relatively low penetration power, and so they cannot be used for

Table 2 Comparison of the three different types of radiation

Property	Alpha particles	Beta particles	Gamma rays
Electric charge	Positively charged ($+2e^-$)	Negatively charged ($+1e^-$)	None
Ionising power	100 times more than beta rays and 10,000 times greater than gamma rays	Comparatively low	Very less ionising power
Penetrating power	Low (can travel 2–4 cm)	Greater than alpha rays (can travel 2–3 m)	Very high penetrating power (can travel 500 m)
Velocity	Relatively slow	Less than the velocity of light	3×10^8 m/s
Mass	Relatively heavy	About 8000 times lighter	None

measurement from outside the body. Because beta particles (electrons) have a modest penetrating strength, they provide good therapeutic effects close to where they are released and may be identified by sensitive counting equipment. Due to their great energy, gamma rays are easily detected by counters, which are outdoor radiation monitoring devices.

Check Points

- Classify the types of radiation.
- State non-ionising radiations give examples.
- Explain the mode of action of ionising radiation.
- Explain the properties of alpha, beta, and gamma rays.

5 The Inverse Square Law

Before proceeding further in this chapter, let us consider what happens as the absorber moves away from the radiation source or what the impact of distance would be on the radiation beam intensity. The radiation is discharged in all directions from a radioactive source. It can be assumed that the source is surrounded by spheres of identical radiation power, with the number of photons or particles decreasing with increase in distance from the source. A property of a radiation beam released from a very tiny source is that it is constantly spreading out or diverging as it goes away from the source; the width of the region covered at any point along the beam is directly proportional to the distance from the source. For example, the cross-sectional beam area is 1 unit wide at a distance of 1 m and develops to a width of 2 units at a distance of 2 m. The space is three units broad at 3 m.

Consider the amount of radiation that traverses the three regions. It is assumed that no radiation is absorbed from the beam before it hits the third area, i.e., the total amount of radiation emitter is the same at all locations and distances from the source. Thus, the radiation intensity, I, depends on the distance from the source according to the Eq. (1):

$$I \propto \frac{1}{r^2} \tag{1}$$

This is called inverse square law. Although the total amount of radiation does not change as it goes away from its source, its concentration decreases. This information is helpful in minimising the dose of radiation given to the patient. The number of disintegrations of a radionuclide per unit of time is called the radioactivity.

If, N_t total number of radio atoms at a time t and dN_t is the number of radio atoms that decay in a small time interval dt, then the radioactivity R_t of the sample is calculated by dividing the number of radioactive atoms that have decayed (dN_t) (Eq. (2)) by the interval dt during which the decay took place.

$$R_t = \frac{-dNt}{dt} \tag{2}$$

A negative sign indicates that the radio atoms are reducing in number.

The metabolic alterations caused by ionising radiation are the primary events contributing to radiation injury in tissues. Therefore, accurate radiation dose measurement is extremely important in any medical application of radiation, both for diagnosis and therapy. The measurement of radiation can be done in a number of ways depending on the quantity, energy associated with it, and kind of radiation.

Actually, the inverse square law helps in calculating the source-to-film distances in X-ray techniques. It is also used to calculate the safe distance from the radiation source.

6 Quantity and Units

There are many types of units to quantify the radiation depending on the feature of α-ray, β, or γ radiation used. The selection of the optimal quantity is dependent on the particular application. The main aim of this chapter is to develop concepts about different quantities and units of radiation. It is also critical to recognise how the units have evolved over time. United States regulatory bodies continue to use traditional units, whereas the majority of the globe and scientific research employ SI units. Both of these are discussed here.

There are two general categories to describe the radiation quantities. The first group consists of quantities that convey the total radiation dose, while the second group consists of measurements that represent the radiation concentration at a specific location.

6.1 Photons

Since gamma rays are high-energy photons, the number of photons can be used to indicate the amount of radiation. However, the number of photons is not frequently employed in practice.

6.2 Fluence (Concentration)

This is the more precise name for the photon concentration of radiation conveyed to the patient. It is defined as the radiation energy delivered to a given region per unit time. The millijoule (mJ) per square centimetre or erg per square centimetre are the units used to express radiation energy concentration.

6.3 Total Photons

Counting the photons that enter the entire exposed area will reveal how much radiation energy has been given to the patient overall. The extent of the exposed area and the radiation dose depend on the total photons. Calculating the total number of photons that will enter the patient if the radiation is evenly distributed throughout the exposed region requires multiplying the concentration (fluence) by the exposed area. The amount of photons and radiation that enters the patient is decreased by reducing the exposed area.

6.4 Energy

The amount of delivered radiation is expressed in energy units. The conventional unit for measuring radiation energy is the electron volt, or eV, which is the kinetic energy obtained by an electron upon acceleration when a potential difference of 1 V is applied and is equivalent to 1.6×10^{-19} J. It is the primary energy unit in nuclear physics.

An eV represents a very small quantity of energy, so it is convenient to use kiloelectron volts (keV) or megaelectron volts (MeV) in several applications. When dealing with particle radiation, the electron volt is a useful unit because it is easy to get the energy gained from an electric field from the product of the potential difference and the number of electric charges the particle possesses. For example, an alpha particle with a charge of +2 will gain 2 keV of energy when accelerated with a potential difference of 1000 V.

Isotopes emitting α-particles have energies in the range of 4.0–8.0 MeV, whereas β and γ emitters typically decompose with energies less than 3.0 MeV.

The Joule (J) is the SI unit of energy. It is related to the electron volt by the formula:

$$1\,\mathrm{eV} = 1.6 \times 10^{-19}\,\mathrm{joules} \quad (3)$$

Or

$$1\,fJ\left(=10^{-15}\,J\right) = 6.241 \times 10^{3}\,\mathrm{eV} \quad (4)$$

6.5 Total Energy

The total energy passed on to a body by a radiation beam is the fluence multiplied by the surface area.

7 Measurement of Radioactivity

Several quantities may be used for the quantification of the radiation. Radiation is determined either in terms of exposure or an absorbed dose. In general, the absorbed dose is regarded as the most accurate method for quantifying irradiation absorption. Scientists use different terms to describe radiation, depending on where it is and what it is doing at the time of measurement.

Figure 4 shows a typical radiation system. It consists of a radiation source, a radiation beam, and a material that absorbs the radiation. Therefore, the quantities associated with all three parts of the system can be measured.

This form of arrangement occurs when the source's radiation is used to irradiate a patient (i.e., the absorber). An imaging device is placed behind the patient for diagnostic purposes, and radiation is designed to cause damage or injury to a specific area of the patient for therapeutic purposes. It is also an example of how absorbers may be working with a radiation source.

7.1 The Radiation Source

The count rate is the most basic measurement. It is the number of interactions between radiation and a detector in a particular time interval. This is expressed in terms of counts per second (cps) or counts per minute (cpm). This is useful for measuring ionising particles, although it can also be used for measuring small amounts of X-rays or gamma rays. Although it is common to express count rate in terms of disintegrations per second (dps) or disintegrations per minute (dpm), this does not give a true count rate.

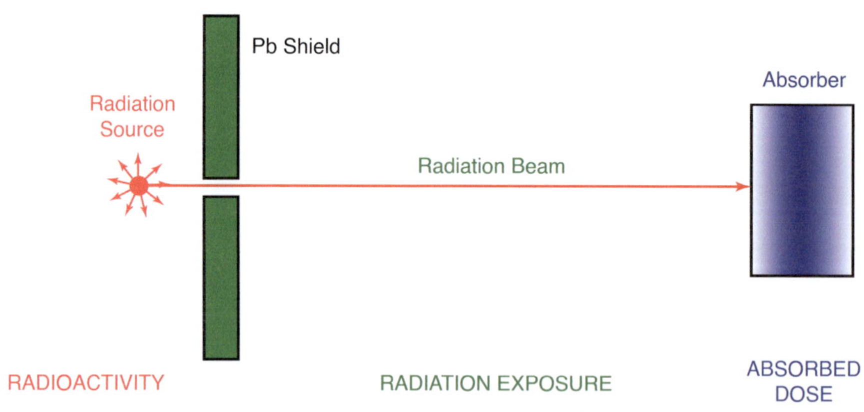

Fig. 4 Radiation system for measurement of radioactivity

When the radiation source is radioactive, the radioactivity of the source is measured. The radioactivity is expressed as the **curie** (traditional unit), the **becquerel** (SI unit), and the Rutherford. These are the units of source strength.

7.1.1 Curie

Curie is defined as the amount of radioactive material in which the rate of disintegration per second is identical to that of radium, i.e.,

$$1\,\mathrm{Ci} = 3.7 \times 10^{10}\,\mathrm{dps} \tag{5}$$

For tracer experiments, in place of curie (which is a large unit), millicurie (mCi) is used.

$$1\,\mathrm{mCi} = 10^{-3}\,\mathrm{Ci} = 3.7 \times 10^{7}\,\mathrm{dps} \tag{6}$$

$$1\,\mu Ci = 10^{-6}\,\mathrm{Ci} = 3.7 \times 10^{4}\,\mathrm{dps} \tag{7}$$

The Curie represents the actual number of disintegrations occurring in a sample rather than the number of disintegrations enumerated in a radiation counter.

7.1.2 Becquerel

Since 1975, the SI system (System International d'Unites, stationed in Paris, France) has adopted becqueral (Bq) as an accepted unit of radioactivity to replace Ci. The Becquerel is equal to one disintegration per second, or 2.703 × 10^{-11} dpm. A curie is thus equal to 3.7 × 10^{10} Bq.i.e.,

$$1\,\mathrm{Ci} = 3.7 \times 10^{10}\,\mathrm{Bq} \tag{8}$$

or

$$1\,\mathrm{Bq} = 2.703 \times 10^{-11}\,\mathrm{Ci} \tag{9}$$

One curie is written as

$$1\,\mathrm{Ci} = 37\,\mathrm{GBq} \tag{10}$$

One millicurie is written as

$$1\,\mathrm{mCi} = 37\,\mathrm{MBq} \tag{11}$$

One micro curie is written as

$$1\,\mu Ci = 37\,\mathrm{kBq} \tag{12}$$

7.1.3 Rutherford (Rd)

The Rutherford is a non-SI radioactive decay unit. It is the unit that measures how many nuclei decay per second occur in radioactive material. It is, therefore, equal to one mega becquerel, and 1 becquerel equals 1 micro Rutherford.

$$1\,\text{Rutherford} = 10^6\,\text{dps} \tag{13}$$

$$3.7\times 10^4\,\text{Rutherford} = 1\,\text{Ci} \tag{14}$$

The source of strength is measured by a GM counter or a scintillation counter, as will be discussed in the chapters to follow.

Various units that are used to measure source strength are listed in Table 3.

By calculating the quantity of radiation, the source of radiation can be identified, which can help in taking the necessary steps to avoid or reduce the exposure.

7.2 The Radiation Beam

Radiation exposure is the property of the radiation beam that is commonly measured. Radiation exposure is often quantified by inserting a tiny volume of air at the point of measurement and then counting the number of ionisations produced. The enclosure for the air volume is an ionisation chamber. When ionising radiation encounters matter (air), i.e., interact with the electrons in the atomic shell, ions are formed. Actually interaction causes the release of electrons from the atom, thus producing an ion pair. Therefore, radiation exposure is expressed in terms of the total ionisation that takes place in the air.

Table 3 Units to measure source strength

S. No.	Unit (abbreviation)	Description and definition
1	Becquerel (SI Unit-Bq)	1 dps or 2.703×10^{-11} dpm
2	Counts per minute (cpm)	Rate of ionisation events
3	Counts per second (cps)	Rate of ionisation events
4	Curie (Ci)	Number of dps equivalent to that for 1 g radium; 3.7×10^{10} dps
5	Disintegrations per minute (dpm)	Actual rate of decay per minute
6	Disintegrations per second (dps)	Actual rate of decay per second
7	Gigabecquerel (SI Unit- GBq)	10^9 dps or 27.027 mCi
8	Megabecquerel (SI Unit-MBq)	10^6 dps or 27.027 μCi
9	Microcurie (μCi)	10^{-6} dps or 2.22×10^6 dpm
10	Millicurie (mCi)	10^{-3} dps or 2.22×10^9 dpm

This quantity indicates the amount of ionisation (i.e., the amount of electric charge) that is caused when a beam travels through air. The electric charge is produced from the formation of ion pairs that reach the detector (that will be discussed later and is expressed as a current). For sources emitting radiation continuously, radiation exposure can be quantified as a rate (exposure per unit time) or as a total incorporated exposure for sources such as X-ray tubes.

7.2.1 Roentgen

The standard unit of radiation exposure is roentgen, named after Wilhelm Roentgen (the man who discovered X-rays) and represented by the letter R. The roentgen is the amount of X-rays or gamma rays (less than 3 MeV) such that the related corpuscular emission per 0.001293 g of air at STP produces in air ions carrying 2.58×10^{-4} coulombs (1 esu) of the quantity of electric charge of either sign.

7.2.2 Coulomb Per Kilogram (C kg^{-1})

This is the SI unit of radiation exposure. It is the amount of X- or gamma rays such that the electrons released during interaction per kilogram of air at STP produce one coulomb of electric charge.

$$1\text{R} = 2.58\,\text{X}\,10^{4}\,\text{coulomb} / \text{kg air} \tag{15}$$

$$1\,\text{C} / \text{kg} = 3876\,\text{R} \tag{16}$$

This unit of radiation exposure is limited to X-rays or gamma rays only.

7.2.3 Gram Roentgen

It is defined as the amount of γ-ray energy converted into kinetic energy of secondary electrons, that corresponds to the energy absorbed by 1 g of air when exposed to 1 roentgen.

The exposure rate, i.e., the exposure per unit time, is more important than the exposure itself. Its units are the C kg^{-1} s^{-1} and the R hr^{-1}. For a specific photon energy, the exposure is proportional to the photon concentration, or fluence. These are related as:

$$\overset{\text{Photon fluence}}{3.1 \times 10^{10}\,\text{photons} / \text{cm}^2} = \overset{\text{Exposure}}{1\,\text{Roentgen}} = \overset{\text{Energy fluence}}{3000\,\text{ergs} / \text{cm}^2} \tag{17}$$

This relationship varies with photon energy, as both the number of photons that will interact and the number of ionisations produced by each interacting photon are energy-dependent.

7.2.4 Air Kerma

This is an extra radiation number that is sometimes used to show how much radiation is going to a particular place in the patient's body. This is the quantity compatible with the SI system. KERMA stands for Kinetic Energy Released per Unit Mass (of Air). It is the quantity of radiation energy, in joules (J), that is actually deposited in or absorbed by 1 kilogram (kg) of air. It is expressed in units of J/kg, which is equivalent to the radiation unit, the gray (Gy) used for the absorbed dose.

7.2.5 Surface Integral Exposure

Exposure is a concentration expressed in roentgens or coulombs per kg, but it does not indicate the total quantity of radiation delivered, or surface integral exposure (SIE), which is based on the exposure and the exposed area size. It is also called an exposure-area product. The integral surface exposure is reported in the standard values of roentgens per square cm (R/cm^2).

The importance of surface integral exposure is that it represents the total radiation conveyed to a patient, whereas exposure indicates only the radiation concentration at a particular location.

7.2.6 Dose Area Product (DAP)

The dose area product is quite comparable to surface integral exposure. The main difference is in the units used. For a uniformly exposed area, the DAP is just the product of the air kerma (in Gy or mGy) and the exposed area (cm^2). The unit of DAP is Gy-cm$^{2.}$. DAP provides an accurate estimate of the total amount of radiation delivered to a patient during a procedure.

8 The Absorber

Up to this point, it has been discussed about the strength of a radioactive source, which is depicted in terms of the number of Becquerels. As long as they are isolated from their environment, radioactive sources pose no biological risk. However, when exposed to radiation, humans (or other biological systems) receive a radiation dose. Therefore, it is necessary to distinguish between the activity of a radioactive source (measured in becquerel) and the radiation dose that may result from the source. The radiation exposure is dependent on the proximity of the source to those who are exposed. The radiation dose is also dependent on the species of radiation (i.e., whether it is alpha, beta, or gamma rays) and the radiation's associated energy. In nuclear imaging procedures, the body absorbs a significant proportion of the radionuclide-emitted energy. The radiation mainly interact with the body's molecules, and each interaction transfers modest amounts of energy. There are typically several such interactions. This results in an increase in temperature because the radiation energy is converted into heat. This heat produced by radiation is an essential to comprehending the concept of *radiation dose*.

There are two things to think about when it comes to absorbed radiation energy: the quantity (concentration) absorbed in specific body regions and the total amount absorbed.

8.1 Radiation Absorbed Dose

When radiation interacts with the absorber, energy is deposited in it. Radiation damage to the tissue (absorber) depends on the radiation dose received, which is known as the absorbed dose, and it is applicable to all forms of radiation. Therefore, the absorbed dose is defined as the quantity that reflects the amount of energy absorbed at a certain location inside the tissue or the amount of radiation energy taken in by the tissue per unit mass. Not every tissue in the path of the beam will absorb the same radiation dose. The dose absorbed by the tissues near the entrance surface will be much greater than for those that lie deeper in the body.

The traditional unit of absorbed dose is radiation absorbed dose (rad). It is defined as the amount of ionising radiation absorbed by the irradiated tissues and is equivalent to 10^{-2} joules (CGS units: 100 ergs) of radiation energy absorbed per kilogram (CGS unit: gram) of material.

9 Gray

Gray (Gy) is the SI unit of radiation absorbed dose, which is named for the renowned radiobiologist LH Gray.

$$1\,\text{Gy} = 1\,\text{J}/\text{kg} \tag{18}$$

The relationship between the rad and gray is:

$$100\,\text{rad} = 1\,\text{Gy}\left(1\,\text{rad} = 1\,\text{cGy}\right) \tag{19}$$

$$1\,\text{rad} = 10^{2}\,\text{J}/\text{kg} = 100\,\text{ergs}/\text{g} = 0.01\text{Gy} \tag{20}$$

In radiation protection, any of the terms—roentgen and rad—are frequently used because their tissue equivalents are almost identical.

Radiation exposure resulted in skin reddening. Earlier, the quantification of radiation was done on the basis of this reddening. This was known as the *skin erythema dose*. This unit was extremely ambiguous because the degree of skin reddening differed from person to person. Another disadvantage was that the reddening did not show immediately after exposure. This dosage unit (together with the associated uncertainty) is still in use for UV radiation. The minimal erythema dose (MED) is the smallest exposure to UV radiation that causes skin reddening.

10 Dose Equivalent (H)

Dose equivalent (H) can be used to show the real or relative effect of radiation on living things. This is the quantity that is usually used to describe the biological effect on people who are exposed to something at work or in the environment. In clinical facilities, personnel exposure is frequently determined and reported as a dosage equivalent.

SI unit for dose equivalent is Sievert (Sv).

$$1\,\mathrm{Sv} = 100\,\mathrm{rem} \tag{21}$$

11 Roentgen Equivalent Man

The unit used for dose equivalent or occupational exposure is the *rem*, i.e., "roentgen equivalent man". It is defined as the dose that is delivered to a man who is exposed to any type of ionising radiation, i.e., the amount of energy deposited by the radiation in body tissues or organs, when human risk factors are present. It is a measurement that correlates the dose of any radiation to its effect on the biological system.

The dose equivalent depends on the absorbed dose (D), the quality factor (Q), and other factors (N) of the particular type of radiation. For most radiation used in diagnostic procedures (X-rays, γ-rays, and β-particles), the value of quality and modifying factors is one. Thus, numerically, the dose equivalent and the absorbed dose are the same.

> **Key Concept**
> The annual radiation dose limit for workers is 0.05 Sieverts (Sv), or 5 rem.

The rem is said to have approximately the same biological effectiveness as one rad of X-rays or γ-rays. Since the rem is a relatively large unit, milli rem is a more convenient way to express most doses.

$$1\,\mathrm{rem} = 1000\,\mathrm{mrem} \tag{22}$$

Table 4 Quality factor values of different radiation

S. No.	Radiation type	Quality factor
1	X-rays, γ-rays, or β-particles	1
2	Neutrons (depends on energy)	2–11
3	Protons (high-energy)	10
4	Alpha particles	20

The dosage is increased by a "quality factor", since not all radiation has the same biological effectiveness (Q). By a factor of three, for instance, a person getting a dose of gamma radiation will sustain significantly less harm than a person receiving the same dose of alpha particles. Therefore, alpha particles are three times more destructive than gamma rays. Therefore, the quality factor of alpha radiation is three. Table 4 illustrates the quality factor for various types of radiation.

12 Roentgen Equivalent Physical

Roentgen equivalent physical (rep) is defined as the dose of any ionising radiation that produce an absorption of 83 ergs of energy per cubic cm of tissue.

$$1\text{R} = 1.1\,\text{rep} \tag{23}$$

Thus in diagnostic radiology

$$1\,\text{R} = 1\,\text{rad} = 1\,\text{rem} \tag{24}$$

$$1\,\text{rem} = 1\,\text{rep} - \text{for}\,\beta\ \ \text{particle.} \tag{25}$$

$$1\,\text{rem} = 0.25\,\text{rep} - \text{for protons, deutrons} \tag{26}$$

$$1\,\text{rem} = 0.1\,\text{rep} - \text{for}\,\alpha\,\text{particle} \tag{27}$$

There are certain other quantities that show how this type of absorbed radiation energy affects living things, like human tissue. These are equivalent dose (H) and effective dose (E).

13 Equivalent Dose (H)

Equivalent dose compares the effectiveness of different types of radiation on biological tissues. The concept of absorbed dose is applicable to all types of materials. However, while evaluating the impact of radiation on biological tissues, it must be taken into consideration that different types of radiation pose damage to different extents. This damage-producing capacity of a particular dose absorbed varies with the type of ionising radiation. For example, alpha particles have more potential to kill cells than beta particles and gamma rays because they are so densely ionising. To put all ionising radiation on the same level in terms of their potential to cause harm, another unit called the equivalent dose is introduced. The equivalent dose is calculated using *radiation weighting factors,* $w_{R,}$ which are estimates of the ionisation capacity of various forms of radiation. It is the product of the absorbed dose from each type of radiation in a tissue and the radiation weighting factor (W_R) for that particular type of radiation, i.e.,

$$\text{Equivalent dose} = \text{Absorbed dose} \times H = w_R D \tag{28}$$

where D is the absorbed dose.

This value is 20 for alpha particles but only 1 for beta particles and gamma rays.

In addition, the risk of radiation is not similar for different body tissues. For example, the risk of radiation for bone surfaces is lower as compared to breast tissues. This can be considered by administering an equivalent dose to each of the major organs and tissues of the body and weighing it using a parameter known as the tissue weighting factor that is connected to the potential for harm for that organ or tissue.

14 Effective Dose (E)

The effective dose is a quantity that takes into account the response of various organs and tissues in the body to the radiation. This quantity is utilised primarily in radiation protection and is intended to compare the risk of stochastic effects resulting from a non-uniform whole-body radiation dose to that of a uniform exposure. The effective dose indicates the health harm. A stochastic effect is a health effect that occurs at random and for which it is believed that the probability of the effect, rather than its severity, is a linear function of dose (for example, developing cancer). Effective dose is becoming an increasingly valuable radiation parameter for communicating relative risk to patients and coworkers.

The effective dose can be determined as

$$\text{Effective dose}\left(\text{Gy}\right) = \text{Absorbed dose} \times w_T \tag{29}$$

where w_T is the tissue weighting factors of different areas and organs.

The effective dose of the body is the sum of the effective dosages of each exposed area. The effective dose is calculated by adding up the product of the equivalent dose (H_T) for each organ or tissue and the tissue weighting factor for that organ or tissue (W_T), added up over all the body organs or tissues:

$$E = \Sigma W_T H_T \tag{30}$$

where the sum (Σ) is for all the types of tissue involved.

There is another complication in measuring the effect of ionising radiation on an individual: the sensitivity of different tissues in the body to radiation damage varies. For example, bone marrow is more susceptible to damage as compared to the skin. Therefore, in such situations, a weighted sum of the equivalent doses for each organ is calculated. The organ weighting factors take into consideration the sensitivity of each organ to damage. Therefore, bone marrow has a greater weighting factor in comparison with the skin. This weighted average of organ doses is called the effective dose. The effective dose will be the same as the equivalent dose if each organ receives the same equivalent dose because the weighting factors for all the body's organs sum up to one. The uniform whole-body dose can therefore be considered the effective dose because it would have the same effect (in terms of the risk of causing injury) as the actual non-uniform dose.

The collective dose is the product of the dose received per person (in Sv) and the number of persons exposed to radiation per year, i.e., man-sieverts per year. Typically, this unit is used for calculations involving population response and protection.

Although exposure, absorbed dose, and dose equivalent—each describe a unique feature of radiation, all these variables also reflect radiation concentration. The factors that connect the three variables for the radiation types utilised in diagnostic techniques have values of about one in soft tissue.

Absorbed dose
(Energy imparted by radiation to standard mass of tissue)
[Gy]

↓↓

Equivalent dose
(Absorbed dose weighted for harmfulness of different types of radiation)
[Sv]

↓↓

Effective dose
(Equivalent dose weighted for susceptibility of different tissues to harm)
[Sv]

Various SI and conventional units of these quantities are given in Table 5.

Table 5 Comparison between SI and conventional units

S. No	Quantity	Symbol	SI unit	Conventional unit	Conversions
1	Radioactivity	A	Becquerel (Bq)	Curie (Ci)	1 mCi = 37 mBq
2	Absorbed dose	D	Gray (Gy)	rad	1 Gy = 100 rad
3	Dose equivalent	H	Sievert (Sv)	rem	1 Sv = 100 rem
4	Exposure	X	Coulomb per kilogram (c/kg)	Roentgen (R)	1C/kg = 3876 R 1 R = 258 μC/kg

Relative Biological Effectiveness

The relative biological effectiveness (RBE) of radiation gives more consideration to specific radiation effects instead of the general risk. Its value depends on what kind of radiation is being used and what kind of biological effects are being looked at. Relative biological effectiveness is the ratio between the dose (in rads) of a radiation that is to be applied and the dose of γ-rays producing the same biological effect, i.e.

$$\mathrm{RBE} = \frac{\mathrm{rem}}{\mathrm{rad}} \tag{31}$$

15 Specific Gamma Ray Constant

Another important radiation unit is the "specific gamma ray constant" of a radioisotope. This quantity represents the exposure rate created by the gamma rays released by a radioisotope. It is actually a combination of the quantities we have just discussed. When dealing with a radioactive source that releases gamma rays, this amount is highly beneficial from a practical standpoint. It is the rate of exposure per unit of activity at a specific distance from the radiation source. Its SI unit is:

$$\mathrm{C\,kg^{1}\,s^{1}\,Bq^{1}\,at\,1\,m} \tag{32}$$

and the conventional unit is

$$\mathrm{R\,hr^{1}\,mCi^{1}\,at\,1\,cm} \tag{33}$$

For measurement of the radioactivity of radioactive materials, some other units are also employed, that are discussed below:

16 Energy Unit

It is the dose conveyed to tissue by ionising radiation in such a way that the energy absorbed by 1 g of tissue is equal to the energy absorbed by 1 g of water exposed to 1 roentgen of γ-radiation.

17 Specific Activity

In radioactive tracer experiments, the radioisotope is attached to the stable isotope. It thus becomes a requirement to express the quantity of radioisotope present per unit mass. This is called the specific activity. Specific activity is defined as the amount of a given radionuclide's activity per unit of atoms. It is usually given in units of Bq/g or Ci/g. It is a characteristic physical property of that radionuclide.

$$\mathrm{A} = \frac{\mathrm{A}}{\mathrm{M}} = \frac{\lambda N}{\mathrm{M}} = \frac{\lambda N_{\mathrm{A}}}{\mathrm{A}} \tag{34}$$

The specific activity of liquids is measured in Bq/l, while the specific activity of air and gases is measured in Bq/m^3. The specific activity of nuclear fallout can be expressed in either Bq/m^2 or Ci/km^2. Both of these quantities are used to explain radioactive contamination. These are related to each other as:

$$1\,\mathrm{Ci}/\mathrm{km}^2 = 37{,}000\,\mathrm{Bq}/\mathrm{m}^2 \tag{35}$$

When a radionuclide is present as a component of a molecule rather than in its elemental form, the idea of specific activity can be applied to the mass of the molecule rather than the element. For example, if Tc radioactivity is present as pertechnatate notation ($^{99m}TcO_4^-$) rather than as a simple element, one can talk about the specific radioactivity of the sample per unit mass of the pertechnatate notation.

In nuclear medicine, knowledge of the specific activity of a sample is important for the following reasons:

- The toxicity of a chemical or drug is always dependent on the amount of the chemical or drug delivered to the patient. Consequently, it is essential to determine the amount of a specific chemical or drug in the radioactive sample.
- The distribution of a radiochemical in a biological system often depends on the amount of carrier present in the sample.
- The specific activity of a radionuclide can also influence the labelling efficiency of a radionuclide compared to a chemical.
- The specific activity of the radionuclides is especially important when choosing them to make therapeutic medicines, immunoassays, and other diagnostic procedures.
- It is also important for assessing radioactivity in certain environments.

Therefore, depending on the application, i.e., the particular aspect taken into consideration, the way of expression will vary, and accordingly, the specific quantity will change.

Check Points

- State the importance of the inverse square law.
- Describe the units used to measure radioactivity.
- Define roentgen.
- Explain what the radiation absorbed dose measure.
- Mention the SI unit of absorbed dose.
- How are the equivalent dose and effective dose calculated? Explain the method to calculate.

In summary, it can be stated that radiation is a form of energy that can either originate from unstable atoms undergoing radioactive decay or be manufactured by machines. Radiation is emitted from its source as energy waves or charged particles. There are various types of radiation, each with distinct properties and effects and these are expressed in different units.

Further Reading

Ahn J, Carson C, Jensen M, Juraku K, Nagasaki S, Tanaka S (eds) (2015) Reflections on the Fukushima Daiichi Nuclear Accident Toward Social-Scientific Literacy and Engineering Resilience. Springer Open. ISBN: 978-3-319-36346-2.

Al−Agwar FA (2020) Research about (Fear from Radiation) IOSR Journal of Dental and Medical Sciences. Series 8. 19 (1) 60-71. e-ISSN: 2279-0853, p-ISSN: 2279-0861. https://doi.org/10.9790/0853-1901086071.

Alajeeli A, Elmahroug Y, Mohammed SH, Trabelsi A (2019) Determination of natural radioactivity and radiological hazards in soil samples: Alhadba and Abuscabh agriculture projects in Libya. Environmental Earth Sciences 78: 194. https://doi.org/10.1007/s12665-019-8213-4.

Al-Musawi TLM (2018) Calibration of a scanning whole body counter for measurement of the activity of gamma emitting radionuclides in the human body. PhD Thesis. School of Engineering Cardiff University UK.

Al-Obaidi AA (2019) Identification and quantification of radioactive impurities in medical radionuclides. PhD thesis. School of Engineering Cardiff University UK.

Bailey RA, Clark HM, Ferris JP, Krause S, Strong RL (2002) 14 - The nuclear environment. Chemistry of the Environment (2nd edn.) 561-686. Science Direct. Academic Press. https://doi.org/10.1016/B978-012073461-0/50061-2.

Bodansky D (1996) Nuclear Energy Principles, Practices, and prospects 2nd edn. Springer-Verlag New York, LLC. ISBN 0-387-20778-3. Springer-Verlag New York, LLC.

Broadman LM, Navalgund YA, Hawkinberry DW (2004) Radiation risk management during fluoroscopy for interventional pain medicine physician. Current Pain and Headache Reports 8: 49– 55. doi: https://doi.org/10.1007/s11916-004-0040-x

Caron DD, Scheiber HN (eds) (2009) The Oceans in the Nuclear Age: Legacies and Risks. Brill Publication. ISBN: 978-90-04-18155-7 (E-Book (PDF)). ISBN: 978-90-04-15675-3 (Hardback).

Guinn VP (2003) Radioactivity. Encyclopedia of Physical Science and Technology (3rd edn.). 661- 674. Science Direct. Academic Press. https://doi.org/10.1016/B0-12-227410-5/00643-8.

Hall EJ, Kereiakes JG (2001) 69 - Effects of Ionizing Radiation on Cells. Cell Physiology Source Book : A Molecular Approach (3rd edn.) 1185-1201. Science Direct. Academic Press. https://doi.org/10.1016/B978-012656976-6/50161-X.

Harrison J (2009) Biokinetics and dosimetric modelling in the estimation of radiation risks from internal emitters. J Radiol Prot 29: A81. https://doi.org/10.1088/0952-4746/29/2A/S06.

Hill DW (1967) Physics applied to anaesthesia chi: Ionising radiations. J Anaesth 39: 335

Hofmann A, Cookie S (eds) (2018) Wilson and Walker's Principles and Techniques of Biochemistry and Molecular Biology [8th edn.] Cambridge University Press

Horvath AL (1986) Conversion Tables of Units in Science & Engineering. McMillan Reference books. Springer. ISBN: 978-1-349-08559-0.

Jamieson B, Martin J (2014) Physics 3901 Intermediate Physics Lab. The University of Winnipeg, Winnipeg, MB, Canada.

Murphy BL, Traub RJ, Gilchrist RL, Mann JC, Munson LH, Carbaugh EH, Baer JL (1983) Fundamentals of health physics for the radiation protection officer. Prepared for the U.S. department of the Army. Pacific Northwest Laboratory. Operated for the U.S. Department of Energy by Battelle Memorial Institute. Battelle.

Nikhilesh S. Bajaj A, Palan CB (2022) Chapter 5 - Understanding OSL radiation dosimetry and its application. Radiation Dosimetry Phosphors Synthesis, Mechanisms, Properties and Analysis.

Noz ME, Maguire GQ (2007) Units associated with radiation protection. Radiation Protection in the Health Sciences. 59-70. World Scientific. https://doi.org/10.1142/9789812770530_0003. ISBN: 978- 981-270-596-9 (hardcover).

Oh SS, Koh SB, Yong SJ (2012) Radon and environmental diseases. Journal of the Korean Medical Association 55(3): 223-229. https://doi.org/10.5124/jkma.2012.55.3.223

Patel S, Dawood A, Pitt Ford T, Whaites E (2007) The potential applications of cone beam computed tomography in the management of endodontic problems. International Endodontic Journal 40: 818– 830. https://doi.org/10.1111/j.1365-2591.2007.01299.x.

Podgoršak EB (2006) Radiation Physics for Medical Physicists. Part of the book series: Biological and Medical Physics, Biomedical Engineering (BIOMEDICAL). Springer ISBN 978-3-540-29471-9.

Podgorsak EB (2006) Chapter 1: Basic Radiation Physics. Review of Radiation Oncology Physics: A Handbook for Teachers and Students. IAEA publication.

Podgorsak EB (2014) Compendium to Radiation Physics for Medical Physicists: 300 Problems and Solutions. Springer. ISBN: 978-3-642-20186-8.

Powsner RA, Palmer MR, Powsner ER (2013) Essentials of Nuclear Medicine Physics and Instrumentation. John Wiley & Sons, Ltd. Print ISBN:9780470905500. Online ISBN:9781118473481. https://doi.org/10.1002/9781118473481.

Plaats GJ (1969) Medical X-Ray Technique. 3rd edn. Philips technical Library. https://doi.org/10.1007/978-1-349-00061-6.

Princi E (2019) 9. Rubber materials and products testing. Rubber. Published by De Gruyter. https://doi.org/10.1515/9783110640328-009.

Radiation Biology: A Handbook for Teachers and Students. Training Course Series No. 42 (2010). International Atomic Energy Agency VIENNA. ISSN 1018-5518

Saha GB (1979) Fundamentals of Nuclear Pharmacy. Springer Science+ Business Media, LLC. ISBN: 978-1-4757-4021-9.

Saha GB (2001) Physics and Radiobiology of Nuclear Medicine. 2nd edn. Springer. ISBN: 978-1-4757-3497-3.

Saha GB (2010) Fundamentals of Nuclear Pharmacy. 6th edn. Springer. ISBN: 978-1-4419-5860-0.

Saha GB (2013) Physics and Radiobiology of Nuclear Medicine. 4th end. Springer. ISBN: 978-1- 4614-4012-3.

Saha GB (2019) Radiation Safety in Nuclear Medicine: A Practical, Concise Guide. Springer. ISBN: 978-3-030-16406-5.

Site Environmental Report 2000 (2001) Prepared by Brookhaven Science Associates, LLC For the U.S. Department of Energy

Slater RJ (2006) Radioisotopes in Molecular Biology Supplement 1. Biochemistry. Wiley Online Library. https://doi.org/10.1002/3527600906.mcb.200400162.

Strohfeldt KA (2015) Essentials of Inorganic Chemistry: For Students of Pharmacy, Pharmaceutical Sciences and Medicinal Chemistry. John Wiley & Sons, Ltd. Print ISBN:9780470665589. Online ISBN:9781118695425. https://doi.org/10.1002/9781118695425.

Wilkinson VM, Gould GW (1998) A. Detection and Measurement of Radioactivity. Experimental Biosciences. Resources for introductory & intermediate level laboratory courses. Food Irradiation A Reference Guide 7-12. Woodhead Publishing Series in Food Science, Technology and Nutrition. https://doi.org/10.1533/9781855736351.5

Woodhead Publishing Series in Electronic and Optical Materials 99-128. Science Direct. Woodhead Publishing. https://doi.org/10.1016/B978-0-323-85471-9.00007-5.

Radioactive Decay Laws

Key Learning Objectives

- Radioactive decay
- Process of alpha decay
- Different types of beta decay
- Gamma decay
- Half life of an isotope

Overview

Radioactive decay involves the release of energy as radiation from all types of sources—naturally occurring as well as artificially produced. During the process, the nucleus tries to attain a stable configuration. The process of radioactive decay of the elements can be achieved by three main methods: by the release of electrons (i.e., beta decay); by releasing helium nuclei (i.e., alpha decay); or by spontaneous splitting into two fragments (i.e., gamma decay). This decay is always associated with the release of high-energy photons, or gamma radiation. The rate of decay of the elements is expressed in the form of the half-life and each radioactive element has a typical half-life associated with it.

1 Radioactive Decay

As discussed in the last chapter, the ratio of protons to neutrons is important in finding out if an element is radioactive or not. An unstable radioactive nucleus can gain stability through the emission of ionising particles and radiation. This process is called "*radioactive decay*". Hence, radioisotopes tend to acquire stability by means of radioactive decay. *Nuclear decay*, or *nuclear disintegration*, or *nuclear*

A. Pathak, *Tools and Techniques in Radiation Biophysics*,
https://doi.org/10.1007/978-981-99-6086-6_4

transformation, or *radioactive decay*, is a numerical phenomenon. The laws that control nuclear decay and the growth of radioactive substances were first composed by Ernest Rutherford and Frederick Soddy in 1902 and then polished by Harry Bateman in 1910.

Key Concept
^{99m}Tc is used as a medical diagnostic tool, especially in imaging techniques, because it has the advantage of being easily attached to various chemical compounds that can be injected into the patient.

This decay starts with the parent nuclide (one type of atom) and converts it to the daughter nuclide (an atom of a different type). Nuclear stability may be attained either in a single step (i.e., by direct decay) leading to the formation of a stable nuclide or by decaying in multiple steps (radioactive series or chain) forming a number of radionuclides and ultimately forming a stable nuclide. For example, the decay of ^{99}Mo or ^{113}Sn undergoes the following steps:

$$^{99}\mathrm{Mo} \rightarrow {}^{99m}\mathrm{Tc} \rightarrow {}^{99}\mathrm{Tc} \rightarrow {}^{99}\mathrm{Ru}(\mathrm{stable}) \tag{1}$$

$$^{113}\mathrm{Sn} \rightarrow {}^{113}\mathrm{In} \rightarrow {}^{113}\mathrm{In}(\mathrm{stable}) \tag{2}$$

The radioactive decay involves a transition from the original nuclide's (parent) quantum state to the product nuclide's (daughter) quantum state. The difference in energy between the two levels involved in a radioactive transition is called the decay energy. The decay energy is released either in the form of electromagnetic radiation (mostly gamma rays) or the kinetic energy of the reaction products.

There are three main processes through which radionuclides tend to attain stability. These are alpha (α), beta (β), and gamma (γ) decay. Naturally occurring radioactive nuclei degrade via α, β, and γ emission, while artificially produced nuclei decay by spontaneous fission, neutron emission, proton emission, and heavy-ion emission.

Check Points
- What happens during radioactive decay?
- What is a radioactive series? Give example.

Thus, it can be summarised that radioactive decay involves the spontaneous alteration of an unstable element into another element with the emission of high-energy radiation. The production of various decay products to attain that balance is known as the decay chain.

1.1 Alpha decay

Alpha (α) decay was the first method of radioactive decay investigated during the 1890s. This type of decay played a very vital role in modern physics experiments that led to the currently accepted Rutherford–Bohr atomic model. Alpha decay is identified by a nuclear transformation in which an unstable parent nucleus attains a more stable nuclear configuration (the daughter) through the release of an α-particle.

Key Concept
Alpha decay is much more easily shielded than other forms of radioactive decay.

For example, the elements with higher atomic weight, decay by the release of α-particles, i.e., a Helium (He) nucleus having two each of neutrons and protons (as already discussed in the previous chapter). Accordingly, the emission of an α-particle results in considerable dimming of the nucleus as it decreases the atomic number (Z) by two units and the atomic mass (A) by 4 units. α-decay can be expressed as:

$$X_z \rightarrow Y_{Z-2} + \alpha_2 \tag{3}$$

For example, radium decay occurs as

$$^{222}\mathrm{Ra}_{88} \rightarrow ^{4}\mathrm{He}_2 + ^{218}\mathrm{Rn} \tag{4}$$

Radium is a familiar example that decays by α-rays; however, this radiation has little biochemical utility, and the radionuclides that emit it are extremely dangerous to life. Another example is the uranium-238 decay that is expressed as:

$$^{238}\mathrm{U}_{92} \rightarrow ^{4}\mathrm{He}_2 + ^{234}\mathrm{Th}_{90} \tag{5}$$

In uranium decay, the ^{238}U nucleus ejects an alpha particle (helium nucleus) and becomes ^{234}Th.

Key Concept
Alpha decay can provide a safe power source for radioisotope thermoelectric generators used for space probes.

Table 1 Some alpha-emitting radioisotope sources

Source	Half -life
^{241}Am	433 years
^{242}Cm	163 days
^{231}Pa	3.2 × 10^{4} years
^{210}Po	138 days
^{218}Po	3 min
^{239}Pu	2.4 × 10^{4} years
^{226}Ra	1600 years
^{222}Rn	3.8 days
^{210}Th	7.5 × 10^{4} years
^{232}Th	1.4 × 10^{10} years
^{234}U	2.45 × 10^{5} years
^{238}U	4.5 × 10^{9} years

Adapted from Gross alpha and beta activities of airborne particulate samples from Wawel Royal Castle Museum in Cracow, Poland. J Radioanal Nucl Chem (2013) 295:1567–1573. By Akbulut et al.

The energetic α-particle that is released slows, down while moving through the medium and captures two electrons from the surrounding medium to become a neutral $^{4}He_2$ atom. The kinetic energies of α-particles produced by naturally occurring radionuclides are between 4 MeV and 9 MeV, corresponding to a range in air of about 1–10 cm, respectively, and in tissue of about 10^{-3} cm and 10^{-2} cm, respectively.

Alpha emitters are not frequently used in biological work. Many of the naturally occurring radionuclides exhibit alpha decay, but alpha-emitting nuclides do not have any medical applications. All elements above lead (Pb) in the periodic table have at least one isotope that decays by emitting alpha particles. Alpha decay is relatively rare because it requires a lot of energy to take two neutrons and two protons out of a nucleus. So it is only possible in some cases. Table 1 illustrates the examples of some alpha-emitting radioisotopes along with their half-lives.

1.1.1 Interaction with Other Atoms

Alpha particles have very high energy, ranging between 4 and 9 MeV. These particles excite the neighbouring atoms either to glow or to be ionised. Since they are large sized and slow-moving with a double positive charge, they very often collide with the atoms coming across their path and thus cause severe ionisation and excitation. For such activity, their energy is quickly dissipated, and so they cannot penetrate deeply. On interaction with matter, α-particles lose only a small amount of their energy. They react with matter in two ways:

- *Excitation*: Here, energy is passed on from the α-particles to the electrons of adjacent atoms these electrons go to higher orbitals. The α-particle pursues its path with its energy decreased by a little more than the amount of energy transferred to the electron. The excited electron ultimately comes back to its original orbital, thereby emitting energy as photons in the visible or near-visible range.
- *Ionisation:* During ionisation, the target electron is completely removed, leading to the formation of the ion pair, consisting of a positively charged ion and an electron.

Thus, alpha decay is a radioactive decay in which a nucleus releases an alpha particle. Alpha decay leads to the formation of another element with two protons and two neutrons less than the original one. It is prevalent in elements with high atomic numbers, like uranium, radium, and thorium.

Check Points
- What is alpha decay? Give example.
- What happens when an alpha particle interacts with matter?
- State the difference between ionisation and excitation.

1.2 Beta Decay

Key Concept
Radioisotopes that decay via beta emission are widely used in science and medicine, particularly in the field of oncology—i.e., cancer diagnosis, follow up evaluation and radio therapy planning.

During this process, a neutron or a proton within the nucleus of a radionuclide is converted into a proton or a neutron, respectively. When a proton is transformed into a neutron, the positive charge within the nucleus decreases by one, thus reducing the repulsion between protons. In contrast, when a neutron is converted into a proton, the positive charge within the nucleus increases by one, resulting in increased repulsion between protons. The consequence of the decrease or increase in the repulsive force in the nucleus is to maintain a balance between the two forces (electromagnetic and strong) and ultimately stabilise the nuclide. Β-decay occurs through one of the following processes:

1.2.1 β^- Emission or Electron Emission, or Negatron Emission

Key Concept
Neutrinos are often called 'ghost particles' because they barely interact with anything else.

In some cases, nuclides with an excess of neutrons may gain stability by converting a neutron to a proton, resulting in the ejection of negative β-particles called negatrons (e^-). An electron is released to balance charge, and to balance energy, a neutrino is also formed, as shown below:

$$n \rightarrow p + \underset{(\text{Negatron})}{\beta^-} + \upsilon \tag{6}$$

where υ is neutrino which can be ignored.

Because neutrinos have an extremely small probability of interaction with matter, they remain undetectable for practical studies. A negatron is actually an electron. But the term negatron is preferentially useful because of the nuclear origin of the particle. As a result of this ejection, the nucleus loses one neutron while gaining one proton; Z, therefore, increases by one, and A remains constant. As a result, the N:Z ratio decreases. The change in atomic number means that the atom has transformed into an element with the next higher atomic number. This process is also known as beta minus decay because the electron is emitted as a beta particle. The resulting nucleus may release its excess energy in the form of gamma rays. The isotope is, therefore, converted into a new element. B-decay can be expressed as:

$$^{A}X_{z} \rightarrow {}^{A}Y_{z+1} + e^- \tag{7}$$

For example, decay of iodine-131 to xenon-131 with the release of an electron:

$$^{131}I_{53} \rightarrow {}^{131}Xe_{54} + e^- \tag{8}$$

The electron is a beta minus particle. In this type of decay, the mass number remains the same while the atomic number increases by one. This is the characteristic feature of this decay. In doing so, a beta particle is released.

Important examples undergoing this type of decay include ^{3}H, ^{14}C, ^{35}S, ^{32}P, etc. (Table 2).

Table 2 Some "pure" beta minus sources

S. No.	Nuclide	Half life	Endpoint energy (MeV)	S. No.	Nuclide	Half-life	Endpoint energy (MeV)
1	^{14}C	5730 years	0.156	7	^{147}Pm	2.62 years	0.224
2	^{45}Ca	165 days	0.252	8	^{35}S	87.9 days	0.167
3	^{3}H	12.24 years	0.0186	9	90Sr/90Y	27.7years/64 h	0.546/2.27
4	^{63}Ni	92 years	0.067	10	^{99}Tc	2.10×10^5	0.292
5	^{32}P	14.3 days	1.710	11	^{204}Tl	3.81 years	0.766
6	^{33}P	25.4 days	0.248				

1.2.2 β⁺ or Positron Emission

The positron is a positively charged β-particle. The nucleus with an excess number of protons achieves stability through the emission of positrons. In this case, a proton is converted into a neutron, followed by the ejection of a high-speed positron.

$$p^+ \rightarrow n^0 + \underset{(\text{Positron})}{\beta^+} + \upsilon \tag{9}$$

Key Concept
The existence of the antiparticle in the electron was suggested by Dirac in 1928.

A neutrino is also formed during this process. Carl D. Anderson named the particle a *positron,* and he was awarded the Nobel Prize for this finding in 1936. Positron emission leads to the loss of one proton and the addition of one neutron, so the atomic number decreases by one while the mass number remains constant. As a result, the N:Z ratio increases. This is identical to ordinary beta decay. But the capacity of the proton and neutron is reversed, i.e., the particle emitted in this case is positively charged instead of negatively charged. Therefore, it is called beta plus decay. The positron emission is represented as:

$$^{A}X_{Z} \rightarrow {}^{A}Y_{Z\text{-}1} + e^{+1} \tag{10}$$

For example, the decay of sodium-22 to neon-22 with the emission of a positron

$$^{21}Na_{11} \rightarrow {}^{22}Ne_{10} + {}^{0}\beta_{+1} \tag{11}$$

Another example is decay of fluorine-18:

$$^{18}\mathrm{F} \rightarrow {}^{18}\mathrm{O} + {}^{0}\beta_{+1} \tag{12}$$

and decay of Iodine-124:

$$^{124}\mathrm{I}_{53} \rightarrow {}^{124}\mathrm{Te}_{52} + {}^{0}\beta_{+1} + \upsilon + \gamma \ \text{ ray} \tag{13}$$

The positron emitted during this decay travels only a short distance before it stops. It will immediately be annihilated by an electron from the surrounding matter, causing both the positron and the electron to disappear. But their energy cannot be destroyed and is converted into two gamma rays, each with 511 keV energy, that are released in opposite directions. This is known as annihilation radiation. Therefore, in this case, the mass and energy of the two particles are converted into two gamma rays released at 180° to each other, and so this phenomenon is also described as back-to-back emission. The γ-detectors are used to detect the positrons.

Key Concept
A PET scan is an imaging technique that helps reveal the metabolic and biochemical functions of tissues and organs.

Positrons, released during β-decay, are used for diagnostic purposes with the help of positron emission tomography (PET). This technique is particularly suitable for locating radioactive isotopes of carbon, nitrogen, and oxygen within patients. For example, the sugar compound fluorodeoxyglucose (FDG) labelled with the radionuclide ^{18}F can be injected intravenously into a patient for positron emission tomography (PET) functional imaging. Based on areas of increased glucose metabolism, the FDG PET scan can detect malignant disease, differentiate benign from malignant disease, be used for staging malignant disease, and monitor response to treatment for malignant disease.

2 Electron Capture

Electron capture was discovered in 1938 by Luiz Alvarez. Since then, it appears that electron capture is a prevalent mode of decay for neutron-deficient nuclides. This is similar to positron emission, where a proton is transformed into a neutron. The number of protons in the nucleus can also be reduced by the process of electron capture. Hence, it yields essentially the same result as positron emission. In this type of decay, one of the electrons from the inner orbital is attracted into the nucleus and is therefore captured by the nucleus. It integrates with the protons to form a neutron; thus, there is a loss of one proton and a gain of one neutron per nucleus. The net result for the atom is to decrease the atomic number by one and increase in the number of neutrons by one. So there is no change in the mass number, which is the sum of the atomic number and the number of neutrons. This is why the atom has converted from one element into another. The neutrino that is emitted has no charge, negligible mass, and a negligible chance of interacting with anything else; therefore, it can be disregarded for most purposes. No particle emission occurs in this case.

$$e^- + p^+ \rightarrow n^0 \tag{14}$$

But X-rays are emitted due to the rearrangement of the energy levels of the electrons. This process is also called "K" capture, as the electron captured by the nucleus is typically from the K-shell, which is the innermost shell. After this occurs, the K-shell will be missing an electron, so other electrons will promptly descend to fill the vacancy, emitting characteristic X-rays as they do so. Similarly, the exchange of a proton for a neutron in the nucleus results in a change in the configuration of nucleons and the emission of the nucleus' excess energy as gamma radiation. For example, ^{125}I decays by electron capture as:

$$^{125}I_{53} \rightarrow {}^{125}Te_{52} + \upsilon + \gamma \text{ rays} + X \text{ rays} \tag{15}$$

Here, ^{125}I has decayed into ^{125}Te, and during this process, energy from the nucleus is emitted as a gamma (γ) ray and an X-ray, as well as some from the atomic electrons.

Iron-55 also decays by this method, being converted into manganese-55.

$$^{55}Fe_{26} + e^{-1} \rightarrow {}^{55}Mn_{25} \tag{16}$$

It is worth mentioning here that the mass number remains unchanged in all types of β-decay while the atomic number is decreased by one. Since the mass number A remains unchanged, β-decay is often called isobaric transition.

In all the above-cited examples, one nuclide has been converted into another nuclide of a different element. The first one is called the parent nuclide, and the newly formed one is called the daughter nuclide. In all these examples, the daughter nuclide formed has been stable so that no more decay takes place. However, this does not always hold true, i.e., sometimes the daughter or progeny nuclide has radioactivity as well, and this can result in multiple generations of radioactive nuclides before a stable nuclide is produced.

For example, strontium-90 (^{90}Sr) is a parent nuclide that decays to a daughter nuclide, yttrium-90 (^{90}Y), by electron emission (i.e., beta minus decay), which is again unstable. This, in turn, also decays by electron emission into zirconium-90 (^{90}Zr), which is a stable species. This can be shown as:

$$\begin{array}{lll} \underset{\text{(Parent)}}{^{90}Sr} \rightarrow & \underset{\text{(Daughter)}}{^{90}Y} & +\text{Beta particle} \\ & \downarrow & \\ & ^{90}Zr & +\text{Beta particle} \end{array} \tag{17}$$

In both negatron and positron emission, the nucleus is left in an excited state. By and large, it can be noted that the isotopes having an excess number of neutrons, like ^{14}C, decay with the emission of negatrons, while the isotopes deficient in neutrons, like ^{11}C, decay with the release of positrons or by electron capture.

3 Interaction with Other Atoms

Negatrons are very small and move rapidly with a single negative charge. They interact with neighbouring atoms and excite the electrons to reach higher orbitals. Then the excited electron loses its acquired energy in the form of photons in or near the visible range of light. By losing energy, it falls back to the ground energy state. This elevation of electrons can cause ionisation if the energy level exceeds the threshold required for the ejection of electrons, thus producing a positively charged cation; if energy the level does not exceed the threshold, the concerned atom only gets excited instead of becoming an ion. The energy of excitation is dissipated as photons, and the atom again comes back to its ground state. Therefore, on interaction with the highly charged beta particles, the molecules may get dissociated, excited, or ionised. But the primary interaction of the particle is with the electrons, and the resulting ionisation is used for the detection of these particles. As the β-particle loses a large part of its energy in a single interaction, the ions produced with its excess energy may produce additional ion pairs. The interaction between β-particles and the nuclei results in scattering or a change in their direction of travel.

Check Points

- State the events that occur during beta decay.
- Define negatron.
- Explain β^- emission with an example.
- What is positron emission?
- State the importance of positrons in radiobiology.
- Explain electron capture in detail.
- How do beta particles interact with matter?

Thus, it can be said that beta decay takes place when an atom has either an excess of protons or neutrons in its nucleus. Out of the various types of beta decay processes, negation emission, i.e., negative beta decay, occurs more frequently than positron emission or positive beta decay.

3.1 Gamma Emission or Isomeric Transition

As already discussed in the previous chapter, protons and neutrons have distinct energy levels in a nucleus, and the nucleus gets excited when either of the nucleons undergoes a transition to a higher vacant energy level. This excited nucleus undergoes a transition to a lower energy level and emits the difference in energy as electromagnetic radiation. This energy is much higher (in MeV) than that during an electronic transition (only a few eV). This emission of electromagnetic radiation from the nucleus is called gamma emission.

In this case, there is no change in atomic number or mass number. Therefore, this type of decay is called an isomeric transition. Also, it involves no change in the element, as was seen in alpha and beta decay. It is just a straightforward transition from an excited state to a ground state. γ emission releases electromagnetic radiation that is similar to X-rays but with a shorter wavelength. There is only a transformation in the nucleus, accompanied by the release of α- and β-particles. A photon carries away some of the energy that is emitted during the process.

It has already been explained that a nucleus can be excited to an excited state (isomer) by the absorption of energy. This is called an *isomer*. It refers to nuclei that share the same atomic number Z and atomic mass number A but have different energy states. Isomers, in general, are of very short duration, except in cases of metastable states. In most α or β decays, within a very short time (i.e., within 10^{-12} s), de-excitation of the daughter nucleus takes place such that the emitted γ-rays seem to be released by the parent nucleus only.

However, in certain other cases, the excited daughter nucleus de-excites not immediately but with a time delay. The excited daughter nuclide is said to be in a *metastable state,* and this process is known as *isomeric transition.* The metastable states have their own half-lives ($t^{1/2}$) (will be discussed later in the chapter) and average lives (τ) and is identified with the letter m next to the atomic mass number, denoting that the nucleus is in a metastable state. For example, $^{137m}B_{56}$ has a half-life of 2.552 min; $^{99m}Tc_{43}$ has a half-life of 6.01 h.

Only gamma rays are emitted during an isomeric transition. Gamma decay can be described as:

$$^{A}X^{*} \rightarrow {}^{A}X + \gamma \tag{18}$$

where * shows an excited or unstable state.

Consequently, isomeric transition occurs via either of the two processes:

Emission of a high-energy photon: the daughter nucleus emits energy as a gamma photon (Fig. 1b); and

Internal conversion, i.e., energy transfer to one of the orbital electrons of the excited nucleus. These two processes are shown below (Fig. 1a):

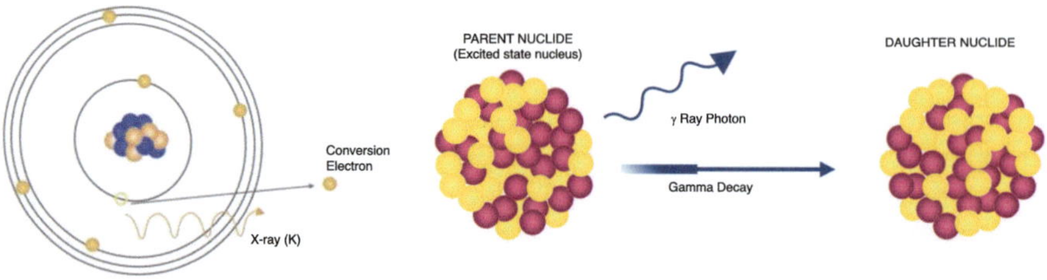

Fig. 1 Isomeric transition occurs through (**a**) Internal conversion and (**b**) Emission of high-energy gamma ray] (Adapted from Nuclear Science and Engineering- an introduction by the Indira Gandhi Centre for Atomic Research (IGCAR), India, supplied by P Swaminathan, Director Electronics and Instrumentation Group, IGCAR, Kalpakkam, July 2005

Check Points

- Explain isomeric transition.
- What is meant by the term "internal conversion"?

Thus, these three types of decay processes help the nucleus attain stability. Any decay process is governed by the same basic law. Gamma decay forms no new element, i.e., no particles are ejected from the nucleus, but the resulting element has less energy because some of the energy is liberated in the form of gamma rays. Gamma rays are the most harmful external hazard. They can penetrate an individual, damaging the cells that come in their path. Beta particles, on the other hand, can partly penetrate skin, thereby producing "beta burns". Alpha particles are not able to infiltrate intact skin.

Key Concept

The radioactive decay law was formulated in 1902 by Ernest Rutherford and Frederick Soddy.

4 Radioactive Decay Law

The disintegration of an unstable nucleus is completely unpredictable and random. The only thing that can be said about an unstable nucleus is that it will decay within a given time period. The proportion of nuclei that decay in a given time for a specific isotope is a constant known as its disintegration constant. If the decrease in the activity of a radioisotope is plotted against time (Fig. 2), a typical exponential form

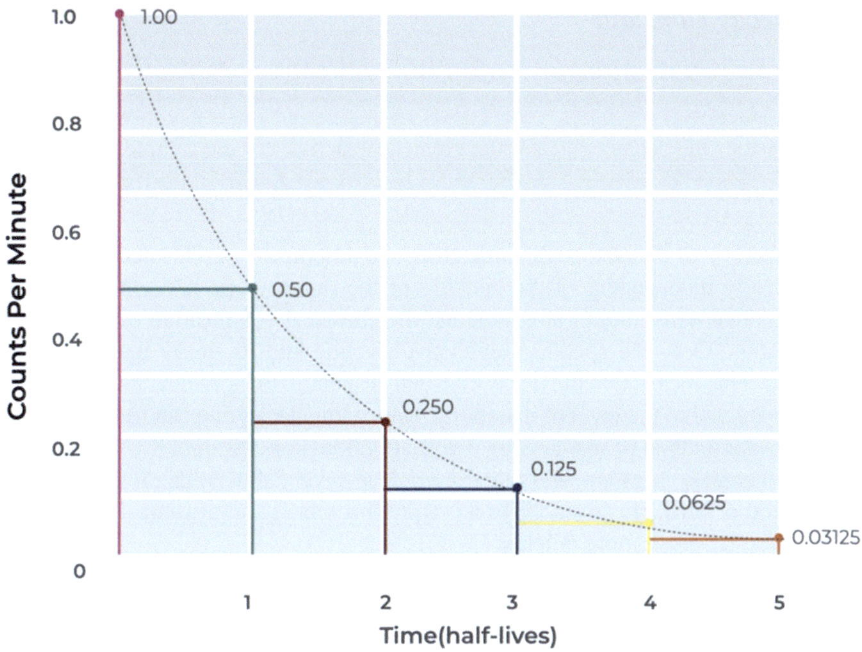

Fig. 2 Disintegration of radioisotopes with time

of the curve is obtained. The total number of atoms that are decaying at any particular time period is directly proportional to the total number of radioactive atoms (N) present at that particular time (t). The exponential curve can be expressed mathematically by the equation:

$$-\frac{dN}{dt} \propto N \tag{19}$$

or

$$-\frac{dN}{dt} = \lambda N \tag{20}$$

or

$$-dN = \lambda N.dt. \tag{21}$$

where $-dN/dt$ = the number of atoms decaying per small interval of time, and λ = decay constant.

The negative sign indicates that the activity is diminishing.

4.1 Decay Constant

Key Concept
The greater the decay constant, the greater the activity of the sample.

It is the fraction of the isotope disintegrating in a unit time (t^{-1}). This decay constant is completely independent of the half-life of the radioisotope and other physical conditions like temperature, pressure, and the chemical composition of the atom's surroundings. Only the specific radioactive nuclide and its decay mechanism is important. This means that the decay constant is specific to individual radionuclides and is connected to the level of instability. Hence, the decay rate can be used as the most relevant method of recognising a given nuclide. For example, ^{238}U has a small value of the decay constant, and therefore it decays very slowly, i.e., it takes a very long period of time. However, ^{99m}Tc has a proportionately large value of the decay constant, so it decays more quickly.

From the curve, it can be seen that if 50% of a sample disintegrates in 1 day, on the second day, 50% of the remaining (i.e., only 25%) will disintegrate. This goes on indefinitely, but the whole substance never disintegrates completely, although its actual amount may become too small to be measured. Therefore, the time of complete disintegration of all radioactive elements is infinity (Fig. 2).

Radioactive decay is an irreversible process for each nuclide.

Dividing by *N* on both sides, Eq. (21) can be written as:

$$-\frac{dN}{N} = \lambda.dt \tag{22}$$

This equation holds good for a brief time interval, *dt*. To determine what occurs over all time periods, the events of each fleeting interval are added. This equation is integrated, taking the limits of time from $t = 0$ to any time t and the number of radioactive nuclei from N_0 to N_t, such that:

$$-\int_{N_0}^{N_t} \frac{dN}{N} = \lambda \int_0^t dt \tag{23}$$

Solving the above equation, we get

$$\ln \frac{N_t}{N_0} = -\lambda t \tag{24}$$

where *Nt* is the number of radioactive atoms present at time *t*, and *No* is the number of radioactive atoms originally present.

The decay constant can be expressed in terms of the half-life ($t_{1/2}$), which is the time required for the activity's value to decline by 50%.

$$\therefore \frac{N_t}{N_0} = \exp(-\lambda t) \tag{25}$$

$$\therefore N_t = N_0 \exp(-\lambda t) \tag{26}$$

This is known as the *Radioactive Decay Law*. It states that the number of radioactive nuclei will decrease exponentially with time, with the decay constant controlling the rate of decrease.

4.2 *Half-life*

The above equation and the radioactive decay law show a correlation between the decay constant and the half-life. For example, when the decay constant is small, the half-life should be long, and correspondingly, when the decay constant is large, the half-life should be short.

Therefore,

$$N_t = \frac{N_0}{2} \tag{27}$$

when

$$t = t_{1/2} \tag{28}$$

Equation (26) can be rewritten by substituting for N and t as:

$$\frac{N_0}{2} = N_0 \exp(-\lambda t_{1/2}) \tag{29}$$

Therefore

$$\frac{1}{2} = \exp(-\lambda t_{1/2}) \tag{30}$$

$$\therefore 2^{-1} = \exp(-\lambda t_{1/2}) \tag{31}$$

$$\therefore \ln 2^{-1} = -\lambda t_{1/2} \tag{32}$$

$$\therefore \ln 2 = \lambda t_{1/2} \tag{33}$$

$$\therefore 0.693 = \lambda t_{1/2} \tag{34}$$

$$t_{1/2} = \frac{0.693}{\lambda} \tag{35}$$

and

$$\lambda = \frac{0.693}{t_{1/2}} \tag{36}$$

The above relationships permit us to compute the half-life of a particular radioisotope if its decay constant is known, and vice versa.

Key Concept
The radioisotopes used in nuclear medicine for diagnostic purposes have short half-lives so that they decay quickly, while the isotopes with longer half-lives will take more time to decay and are therefore suitable for therapeutic purposes.

The half-life of radionuclides may vary from a fraction of a second to billions of years. a value characteristic of the particular isotope. Table 3 displays examples of the half-lives of certain radioisotopes. When several different isotopes of the same element are available for experimental application, the half-life is one of the most important factors in deciding which to use. It can be noticed that some of these have a relatively short half-life. These are the ones that are used for medical diagnostic purposes because they do not remain radioactive for very long following administration to a patient and hence result in a relatively low radiation dose.

Table 3 Half- lives of some radionuclides

S. No.	Radioisotope	Half life (approx.)
1	^{14}C	5730 years
2	^{51}Cr	27.7 days
3	^{137}Cs	30.07 years
4	^{3}H	12.5 years
5	^{131}I	8 days
6	^{81m}Kr	13 s
7	^{32}P	14.2 days
8	^{212}Pb	3×10^{-7} s
9	^{239}Pu	24,000 years
10	^{226}Ra	1600 years
11	^{99m}Tc	6.04 h
12	^{238}U	4.51×10^{9} years

Others have very prolonged half-lives. Example includes: ^{226}Ra has a half-life of more than 1,500 years. These isotopes have previously been used for a number of therapeutic applications.

There are several other terms that are also related to the half-life and are of importance in radiation biology. These are discussed below:

4.2.1 Average Life

In a radioactive decay, all radioactive atoms do not decay at the same rate. Each individual radioisotope survives for a specific period of time. Therefore, average life is the mean of all these durations for which radioisotopes remain in existence.

This is connected to the half-life of the radionuclide and its decay constant as follows:

$$T_{\mathrm{AV}} = 1.44 \times T_{1/2} = 1/\lambda \tag{37}$$

4.2.2 Biological Half-life

The radioisotopes that are ingested or otherwise absorbed will be progressively eliminated by the kidneys, bowel, respiration, and perspiration. This indicates that a radioactive atom *can* be formed prior to its decay. The time elapsed before half of a compound (whether radioactive or not) has been removed by biological means is referred to as the *biological half-life* and is typically denoted as t_b.

4.2.3 Effective Half-life

In nuclear medicine studies, one is more interested in the diffusion and disappearance of radioactive substances or drugs from the biological systems with time because of both physical decay and biological elimination (i.e., via metabolism, diffusion, or excretion). Such information is important for calculating the radiation dosage to be given to a patient and determining the optimal time for figuring out when imaging should be done after radiopharmaceutical administration.

If a radioactive molecule with a physical half-life ($t_{1/2(p)}$) is eliminated from the body with a biological half-life ($t_{1/2(\mathrm{b})}$), the "*effective*" *half-life* ($t_{1/2(\mathrm{eff})}$) is given by the expression:

$$\lambda + \lambda_b = \lambda_{\mathrm{eff}} \tag{38}$$

where λ_{eff} is the effective disappearance probability for a radioactive substance or drug.

The biological half-life is rather uncertain as compared to the exact value of the physical half-life. It is uncertain because the clearance from the body depends upon various factors like sex, age, and the chemical form of the radioactive substance. The biological half-life will vary from one type of animal to another and from one type of plant to another.

If we convert the various disappearance constants in the above equation to their respective half-lives, the equation becomes

$$0.693 / T_{1/2(\mathrm{p})} + 0.693 / T_{1/2(\mathrm{Biol})} = 0.693 / T_{1/2(\mathrm{eff})} \tag{39}$$

or

$$1 / T_{1/2(\mathrm{p})} + 1 / T_{1/2(\mathrm{Biol})} = 1 / T_{1/2(\mathrm{eff})} \tag{40}$$

From the above equation, the effective half-life of a radiopharmaceutical can be determined if the physical and biological half-lives are known. However, $T_{1/2(\mathrm{eff})}$ will always be smaller than or equal to $T_{1/2(\mathrm{p})}$ and $T_{1/2(\mathrm{b})}$.

4.2.4 Radio-Ecological Half-life

It is defined as the radioactive half-life of the animals and vegetation residing in a particular area. Radio- ecological half-lives are even less accurate than physical and biological half-lives. Consider an area contaminated with a radioactive isotope (for example, ^{137}Cs). A portion of the activity will eventually seep into the earth and the water table. Each year, a portion of the activity will be absorbed by the plants, which will then be consumed by the local animal in the area. It differs for various animals and vegetation species.

Check Points

- Describe radioactive decay law.
- What is decay constant? Write its importance.
- Define half-life. Explain its significance.
- Define average, biological and effective half- life.

Thus, the half-life of an isotope is used to characterise the rate at which the isotope decays and emits radiation. After a specified period of time, it is possible to predict the amount of radioactive material that will remain. It is a defining property of the nucleus.

Further Reading

Akbulut S, Barbara Krupinska, Anna Worobiec, Ugur Cevik, Halim Taskın, Rene Van Grieken, Lucyna Samek, Ewa Wiłkojc. Gross alpha and beta activities of airborne particulate samples from Wawel Royal Castle Museum in Cracow, Poland. J Radioanal Nucl Chem 2013; 295: 1567–1573. https://doi.org/10.1007/s10967-012-1983-8

Bailey DL, Humm JL, Todd-Pokropek A, van Aswegen A (2014) Nuclear medicine physics: a handbook for teachers and students. International Atomic Energy Agency

Henriksen T, Maillie HD (2003) Radiation and health. Taylor & Francis

L'Annunziata MF (2007) Chapter 8: Radionuclide decay, mass, and radioactivity units. In: Radioactivity introduction and history. Elsevier, p. 529-552. https://doi.org/10.1016/B978-044452715-8.50011-6.

L'Annunziata MF (2012) Chapter 1: Radiation physics and radionuclide decay. In: Handbook of radioactivity analysis, 3rd ed. Elsevier, p. 1-162. https://doi.org/10.1016/B978-0-12-384873-4.00001-3.

L'Annunziata MF (2020) Chapter 1: The atomic nucleus, nuclear radiation, and the interaction of radiation with matter. In: Handbook of radioactivity analysis, 4th ed. vol. 1. Academic Press, p. 1–243. https://doi.org/10.1016/B978-0-12-814397-1.00001-7.

Petrich T, Helmeke H-J, Meyer G, Knapp W, Pötter E (2002) Establishment of radioactive astatine and iodine uptake in cancer cell lines expressing the human sodium/iodide symporter. European Journal of Nuclear Medicine and Molecular Imaging 29: 842–854. https://doi.org/10.1007/s00259-002-0784-7.

Podgorsak EB (2005) Chapter 1: Basic radiation physics. In: Radiation oncology physics: a handbook for teachers and students. IAEA Publication.

Podgorsak EB (2006) Radiation physics for medical physicist. In: Biological and medical physics, biomedical engineering. Springer

Podgorsak EB (2009) Chapter modes of radioactive decay. In: Radiation physics for medical physicists, 2nd ed. Springer, p. 475–521. https://doi.org/10.1007/978-3-642-00875-7_11.

Podgorsak EB (2010) Radiation physics for medical physicists. 2nd ed. Springer

Podgorsak EB (2014) Compendium to radiation physics for medical physicists. 300 problems and solutions. Springer

Podgorsak EB (2016) Chapter radioactivity. Radiation physics for medical physicists. Springer, Cham, p. 262–358. https://doi.org/10.1007/3-540-29471-6_8.

Stahl A, Schachoff S, Beer A, Winter A, Wester HJ, Scheidhauer K, Schwaiger M, Wolf I. [111In]DOTATOC as a dosimetric substitute for kidney dosimetry during [90Y]DOTATOC therapy: results and evaluation of a combined gamma camera/probe approach. European Journal of Nuclear Medicine and Molecular Imaging 33: 1328–1336 (2006). https://doi.org/10.1007/s00259-006-0078-6.

Valk PE, Delbeke D, Bailey DL, Townsend DW, Maisey MN (eds) (2006) Positron emission tomography clinical practice. Springer

Interaction of Radiation with Matter

Key Learning Objectives

- Interaction of alpha particles with matter
- Concepts of ionisation and excitation
- Interaction of beta particles with matter
- Gamma ray interaction
- Photoelectric effect
- Compton effect
- Pair production

Overview

Different types of radiation interact with matter in a variety of ways. As already discussed, they can be charged particles, i.e., alpha particles, protons, electrons, and positrons, or they can also be uncharged, for example, neutrons, X-rays, and gamma rays. When these radiations interact with matter, they produce a number of effects through various processes. The nature of interactions will depend on the radiation type, energy, and type of material. Penetrating radiation perceive matter as a collection of electrons and nuclei (and their constituents). The charged particles interact continuously via coulombic forces with the electrons in the material through which they travel, and energy is emitted in the form of rays or fast-moving particles. However, uncharged particles are not subjected to Coulomb forces. Actually, they are indirectly ionising, i.e., they first have to produce charged particles.

A. Pathak, *Tools and Techniques in Radiation Biophysics*, https://doi.org/10.1007/978-981-99-6086-6_5

As discussed in previous chapters, alpha, beta, and gamma rays are ionising radiation. As these radiation travel through matter, they transfer some energy to the substance and result in either ionisation (i.e., electrons being knocked out of the atom) or excitation (electrons undergo a transition to higher energy states). This effect can be detrimental (by damaging live cells), or it might provide a valuable basis for detecting the radiation. In order to fully comprehend the benefits and risks associated with these radiation, it is crucial to understand the myriad ways in which they interact.

1 Interaction of Alpha Particles

Alpha particles primarily interact with the electrons of the absorber atoms in matter via Coulomb forces. These particles have an effect on matter mainly via coulomb forces between their positive charge and the negative charge of electrons in atomic orbitals. After penetrating the medium, the charged particle interacts simultaneously with multiple electrons. Alpha and beta particles transfer energy mainly by:

- Excitation: This process occurs when the charged particle's energy is transferred to the atom and electrons are raised to higher energy states.
- Ionisation: This process takes place when the charged particle possesses sufficient energy to eject an electron from its orbit, due to which ion pairs are formed in matter.

Because alpha particles are heavier and doubly charged, they cause extensive ionisation as they strike electrons in the matter, ejecting them out. Because they are heavier than electrons, their path is not much affected, and they continue travelling in a straight path. However, during each collision, a very small amount of energy of the alpha particle is lost, causing it to slow down steadily. The number of ion pairs formed per unit of travel distance is known as specific ionisation. The specific ionisation of alpha particles depends on the energy associated with the radiation. Alpha particles have high specific ionisation. As it loses its energy, it slows down and consequently spends more time near atoms. This is the reason the specific ionisation increased near the end of the travel of alpha particle. Sometimes an alpha particle collides with the nucleus, but it is a rare event due to much smaller size of the the nucleus. However, if alpha particle collides with a nucleus, it will be deflected from its forward trajectory and the nucleus will recoil in different directions (Fig. 1). This recoiled nucleus then causes additional ionisation in the medium.

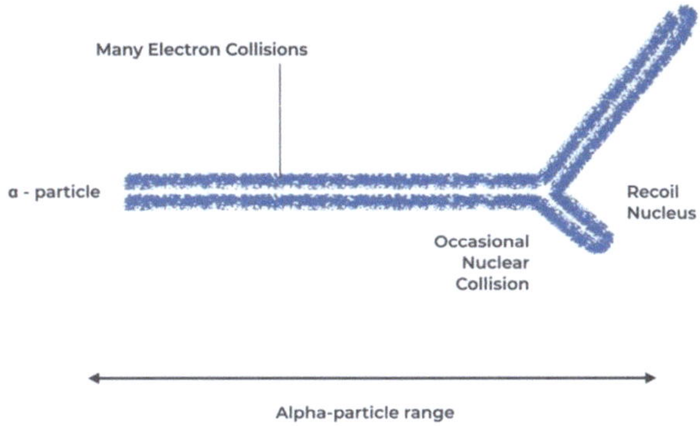

Fig. 1 Alpha particle interaction with matter

> **Key Concept**
> Alpha particles cannot be used for radiotherapy due to their low penetration property.

Since the ionisation produced is very intense, the alpha particle will soon run out of energy due to multiple collisions and ultimately stop. Actually, near the end of the travel, the specific ionisation of the alpha particle decreases to zero, and the particle becomes a neutral atom. The distance moved by the alpha particle before it ultimately stops is the particle's range. The particle range depends on its energy and the medium through which it moves. However, for an alpha particle, every time it is very short. In air, the range of an alpha particle having 1 MeV energy is 5 mm, while in water it is only 7 microns. It is clear and certain that the source of an alpha particle, if present on the outer side of the body, will not cause any harm, as the outer layers of the skin will capture all the alpha particles. However, if it was allowed to go inside, a small number of cells would absorb its radiation and produce very harmful effects. This is why alpha emitters, like ^{238}Pu, are very dangerous if taken inside by some means, and therefore, alpha emitters are not useful in medicine.

> **Check Points**
> - What happens when alpha particles interact with matter?
> - What is a recoil nucleus?
> - State the events that occur when alpha particles transfer energy to atoms.
> - What is meant by specific ionisation?
> - Define the range of the particle.

Thus, the interaction of alpha particles with matter causes the transfer of their energy to the material through which they travel. Due to their low penetrating power, alpha particles are not preferred for radiotherapy. Moreover, if these particles enter the body, they produce a very damaging effect on the internal tissues.

2 Interaction of Beta Particles

Beta particles interact with electrons through the electric fields surrounding these charged particles, which leads to either excitation or ionisation, as seen in the case of alpha particles. Because beta particles are lighter and have only a single charge, less ionisation is produced as compared to alpha particles. Also, these are easily deviated as they ionise the atoms of the material through which they traverse. Often, the collisions with an electron are so high that the beta particle undergoes a large angle deviation and then the electron with which it strikes possesses enough energy to move off. This electron is known as a delta ray, and it moves on to cause further ionisation. Sometimes, if a beta particle encounters a nucleus in a medium with high atomic numbers, it will intensely deviate from its path, and during this process, it emits bremsstrahlung X-rays (Fig. 2). After an extensively snaky track, these beta particles will ultimately stop. Similar to alpha particles, beta particles also have a specific range. However, because they are less ionising, they will move more slowly than alpha particles and therefore have a greater range. For example, a beta particle with an energy of 600 keV moves 2.5 m in air, and in tissue, its range is 3 mm. As all beta sources release a range of beta particles with varying energies, the range of particles will always be diverse. Therefore, the number of beta particles swiftly decreases with the thickness of the material traversed, till all their energy is lost. If a beta source lies near or inside the body, within millimetres, all its radiation will be caught up. This means all the energy is taken up by local tissues, and because beta particles are mildly ionising, there is a risk of tissue damage.

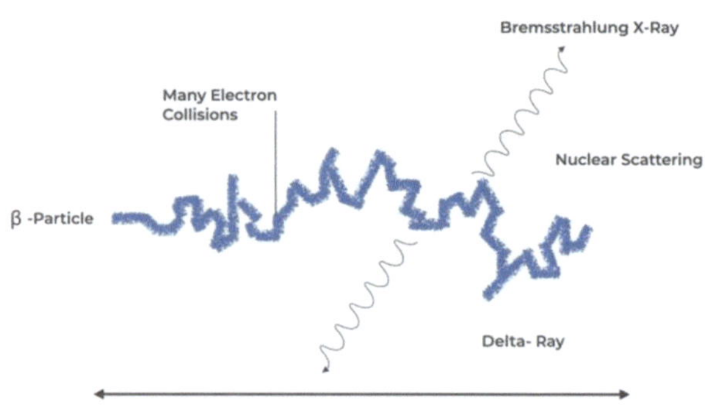

Fig. 2 Beta particle interaction with matter

Key Concept
Bremsstrahlung is produced by a sudden slowing down or deflection of charged particles travelling through matter in the vicinity of the powerful electric fields of atomic nuclei.

In diagnosis, this may be used to minimise the radiation dose and hence the associated hazards, but it is more useful in therapy. For instance, beta particles with significant energy released by the disintegration of ^{90}Sr can be utilised to administer therapeutic dosages to surface tissues. Similarly, the beta particles released by ^{131}I are used in thyroid gland therapy because iodine is known to concentrate more in the thyroid tissue than in other organs. Another example is to use ^{3}H, a low-energy beta particle emitter, in small amounts so that it may be safely delivered inside for diagnostic studies performed *in vivo*. Since beta particles remain in the patient, blood or urine samples must be collected and counted in the laboratory in order to determine the level of activity present. Even then, it is not simple to find low-energy beta particles. Therefore, samples must be thoroughly blended with the medium and detected by a liquid scintillation counter. The beta particles having a slightly higher energy, like ^{35}S, are used in autoradiographic studies.

Key Concept
Beta particles can be used to treat a variety of health conditions, including cancers of the eye and bone.

Check Points
- Why do beta particles produce less ionisation than alpha particles?
- What are Bremsstrahlung X-rays?
- State the uses of beta particles in therapy.
- How are the beta particles measured, and why?

Thus, it can be summarised that beta particles follow zig-zag path while passing through the matter, resulting in a longer range in the matter. They also cause ionisation or excitation, as was seen in the interaction of alpha particles. Bremsstrahlung radiation is produced during this process, and some energy from beta particles is lost and is emitted as X-rays.

3 Interaction of Gamma Rays

Unlike other types of particles discussed above, gamma rays are electromagnetic radiation (or high-energy photons), and therefore they interact with matter in a different way. They lack an electric charge. Therefore, inelastic collisions with atomic electrons are impossible, which are specific properties of charged particles. These rays are several times more penetrating in a medium than charged particles. Also, each gamma ray experiences one or two interactions, while many gamma rays traverse through the medium without any interaction at all; alpha and beta particles, however, undergo many individual interactions while passing through the medium. Gamma rays produce indirect ionisation by first producing secondary electrons, and then these secondary electrons cause further ionisation of the medium, similar to a beta particle. Although alpha particles and beta rays are stopped after several interactions, gamma rays undergo only a few interactions and move forward for a long distance before dissipating their energy.

Key Concept
Paul Villard, a French chemist and physicist, discovered gamma radiation in 1990, while studying radiation emitted from radium.

Thus, the extent of gamma rays is ambiguous. However, the intensity of a beam of gamma rays is reduced by both scattering and absorption operations, such that it decreases proportionally with distance. The half-value layer (HVL) is the distance needed to reduce the number of gamma rays by half. The same exponential mathematics applies to HVL as to the half-life of radionuclide decay. In Fig. 3, when gamma rays pass over the medium's one-half-value layer, half of the gamma rays are either absorbed or scattered, resulting in a beam of low intensity with only 50% of these gamma rays. Another half of the gamma rays would disappear in the next half-value layer, and so on. This value varies with the energy of the gamma rays and the properties of the medium. For instance, 140 keV of a gamma ray has an HVL of 0.25 mm in lead. This means that lead shielding of 0.25 mm will decrease the intensity of gamma radiation to half of the original; 0.5 mm (2 HVLs) will decrease the energy to one quarter; and so on. In tissue, however, the HVL is much higher (i.e., 47 mm) for gamma rays of 140 keV energy. Therefore, if a radionuclide emitting 140 keV gamma rays is present inside an individual, enough gamma radiation will enter body tissues, thereby permitting radionuclide location detection for diagnostic imaging. Gamma rays that remain inside will disperse their energy to various organs, thus leading to a dispersed radiation dose that is far less harmful than the site-specific doses released by beta emitters. That is why gamma emitters like ^{99m}Tc

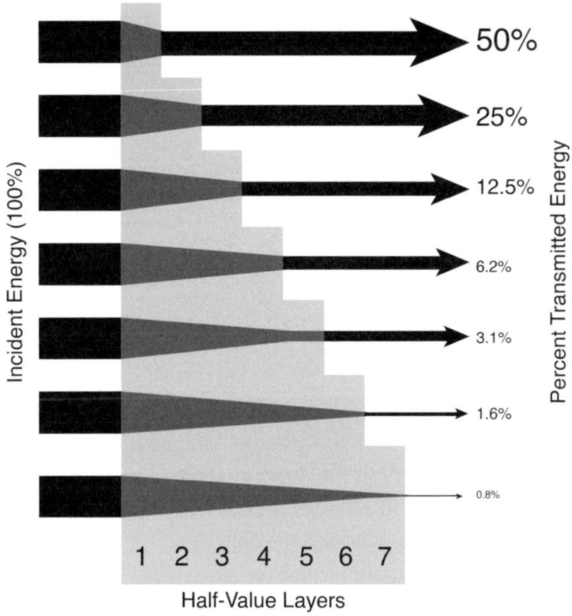

Fig. 3 Gamma ray interaction with matter

are very helpful for diagnostic imaging. Gamma rays produced from ^{60}Co can be used to sterilise many disposable items, such as disposable plastic syringes and needles.

> **Check Points**
> - What are the effects produced when gamma rays interact with matter?
> - Define a half-value layer.
> - What is the reason for scattering of gamma rays?
> - Why are gamma rays preferred for diagnostic imaging?

Thus gamma rays are highly penetrating and interact with matter through ionisation. They are, however, less ionising than alpha particles. Each gamma ray photon is removed individually from the incident beam in a single event.

Gamma rays interaction with the medium occurs in three principal ways: one controlling the other, depending on the gamma ray energy. These three processes are:

- Photoelectric effect,
- Compton effect, and.
- Pair production.

4 Photoelectric Effect

The photoelectric effect refers to the interaction of a photon with a securely bound electron of an absorber atom. This process takes place when low-energy γ rays interact with the electrons in the inner shell of an atom of the material and completely transfer their whole energy to the inner electrons, which get ejected as a photoelectron from the atom (Fig. 4). The incoming photon uses all of the energy in ejecting the electron from the parent atom.

Thus, the atom absorbs the energy of gamma rays and causes ionisation. The photoelectron that is released travels a particular distance within the medium and causes secondary ionisations, and ultimately loses its energy. In this manner, the whole energy of the gamma photon is delivered to the irradiated tissue. This energy transfer occurs in two steps. The first step involves photoelectric interaction, in which the energy of the photon is transferred to the electron, and the electrons deposit this energy in the nearby medium in the second step. Actually, the energy of the photon splits into two fractions during the interaction: some energy is employed to overcome the binding energy of the electron, and ultimately the electron is removed from the atom; and the remaining part of energy is given to the electron in the form of kinetic energy and gets deposited close to the site of interaction.

The photoelectron acquires kinetic energy (E_e) equal to the difference between the photon energy (E_γ) and the electron's binding energy in a particular shell (i.e., $E_e = E_\gamma$ minus the electron's binding energy). Photoelectric interactions mainly take place with electrons that are bound firmly to the atom, i.e., have a high binding

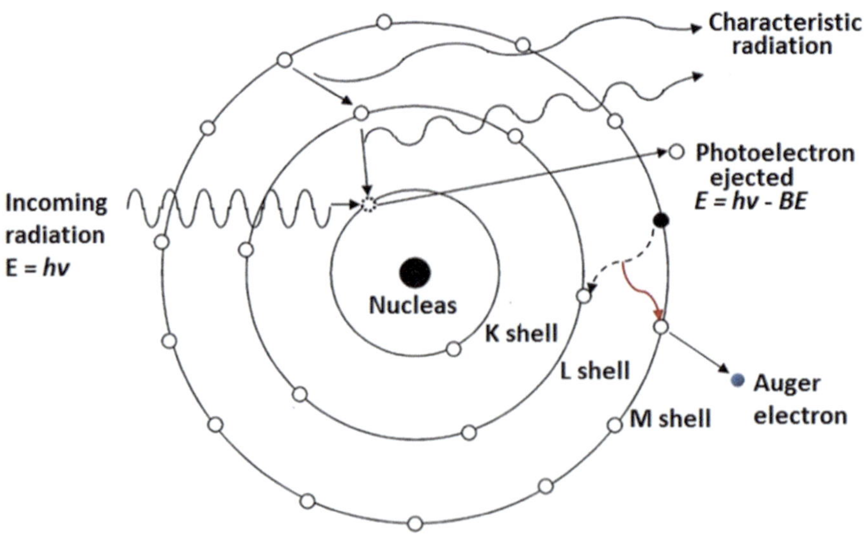

Fig. 4 Photoelectric effect. (*Adapted from Monte Carlo evaluation of the dose perturbation effect of hip prostheses for megavoltage photon radiotherapy, 2015.* Physica Medica 31(1):S7)

energy. Photoelectric interactions are more likely when the electron's binding energy is slightly less than the photon's energy. If the binding energy is greater, no photoelectric interaction takes place. Under such conditions, the electron is removed from the atomic shell. If the electron is removed from some inner shell (e.g., the K shell), a vacancy is created in the shell, which is ultimately filled by an electron in the outer shells (like the L or M shells). A characteristic *X*-ray or Auger electron is released during this process. The energy of this X-ray is dependent on the electron's binding energy. On the other hand, if the gamma ray strikes the outer shell electrons, then the atom gets ionised. Therefore, the chance of the atom undergoing photoelectric interaction in an atom depends on the energy of the gamma photon ($E\gamma$) and the atomic number of the concerned atom.

Key Concept
Gamma rays are extensively used in medicine, especially in the area of oncology, to treat malignant tumours by a process called gamma knife surgery based on the photoelectric effect.

The energy characteristics of the gamma radiation emitted by excited nuclei vary depending on the nuclides involved. Thus, the energy of the γ photons can be regarded as "finger prints" to recognise the nuclides. The absorption of gamma photons by shielding substances and radiation detectors is also due to the photoelectric effect.

The photoelectric effect is the main energy transfer method for gamma ray photons with energies less than 50 keV in the tissues, but it is of less importance at higher energies.

In diagnostic radiology, the primary mode of interaction with the medium is the photoelectric effect. The contrast effect is also caused by it. In therapeutic applications, low energy beams in ortho voltage irradiation caused excessive absorption of energy in the bone.

Check Points
- What is a photoelectric effect?
- When are photons produced, and what is their fate?
- What type of radiation is produced during the photoelectric effect?
- What are auger electrons?

5 Compton Effect (Compton Scattering)

The Compton effect, which is caused by medium-energy γ rays, consists of a γ photon that strikes the outer shell electron and has binding energy that is extremely less as compared to that of an incident photon. During Compton scattering, the incident photon delivers a portion of the energy of the electron, which gets ejected from an atom, as shown in Fig. 5:

> **Key Concept**
> In 1927, Arthur Holly Compton was awarded the Nobel Prize in Physics for discovering the Compton effect in 1923, which proved the particle nature of electromagnetic radiation.

During collision, a new photon of low energy (a radiation of lower frequency) is produced, which proceeds in an altered (deflected) direction. Because of the change in the direction of the photon, this type of interaction is referred to as a scattering process. Again, the ejected electron behaves as a negatron. The direction of this Compton photon (or a secondary photon) is not along the same path as the earlier incoming photon. The angle between the incident photon's direction and the scattered photon's direction is called the scattering angle (θ) and is in the range from $\theta = 0°$ (in case of forward scattering) to 90° (in case of side scattering) and finally to $\theta = 180°$ (back scattering). The exact quantity of energy received by an electron or energy dispersed by the photon is dependent on the scattering angle and energy associated with the incident photon (E**γ**). The larger the scattering angle of a photon, the more energy is lost to an electron. Maximum energy transfer from the photon to the electron takes place when the photon gets scattered during backscattering (i.e., at 180°).

The energy of this back-scattered photon is related to the incident photon energy (Eγ) by the formula:

$$E\gamma \text{ mini} = \frac{E\gamma}{1 + 4E\gamma} \tag{1}$$

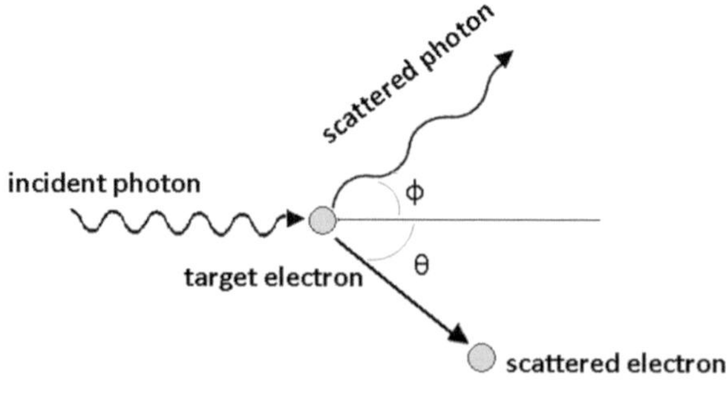

Fig. 5 Compton Effect. (Vinay Venugopal* and Piyush S Bhagdikar. Physics Education 2013, 29(1). Article no 3)

Compton scattering, thus, causes ionisation of the absorbing medium because of electron loss. The scattered electron traverses a certain distance in the medium and ultimately loses energy because of ionisation and excitation in the medium. The probability of the Compton scattering process decreases with an increase in the energy of the photon. It is the main absorption process for γ rays in the intermediate energy range (100 keV– 10 MeV). This range is useful for therapeutic applications. Most of the γ radiation present in a nuclear explosion also has this range. Compton scattering does not depend on the atomic number of the medium, because dense materials, such as lead, are marginally superior shields per unit weight to the less dense ones.

Check Points

- What is Compton scattering?
- What type of radiation is produced during Compton scattering?
- What are back-scattered electrons? When will they be produced?

Thus, the Compton effect helps in identifying unidentified nuclides. However, the Compton scattered electron can mask lower energy photons interacting via the photoelectric effect during detection, making explanation difficult.

6 Pair Production

The pair production process occurs with gamma energies greater than 1.02 MeV and over 5 MeV of energies, this has become very important as an absorption method. In pair production, a positron and an electron are produced when a high-energy gamma ray strikes the nucleus of a heavy atom. For pair production to take place, the γ ray energy must be 1.02 MeV. This energy is then converted into the masses of electrons, (e^-) and positrons (or anti-electrons (e^+)). These particles are released in opposite directions, and they cause damage in the form of secondary charge particles. The positive charge of a positron is equal to the negative charge of an electron, despite having the same mass as that of an electron. These energetic positrons proceed through the medium, undergo collisions with orbital electrons, and lose their kinetic energy via Coulomb interactions with absorber nuclei. The positron has a very short lifespan and ultimately collides with the released electron, and then the two species annihilate. The total mass of these two particles is then split into two oppositely moving γ photons with 0.51 MeV each. This process is known as pair production and is shown in Fig. 6.

Fig. 6 Pair production

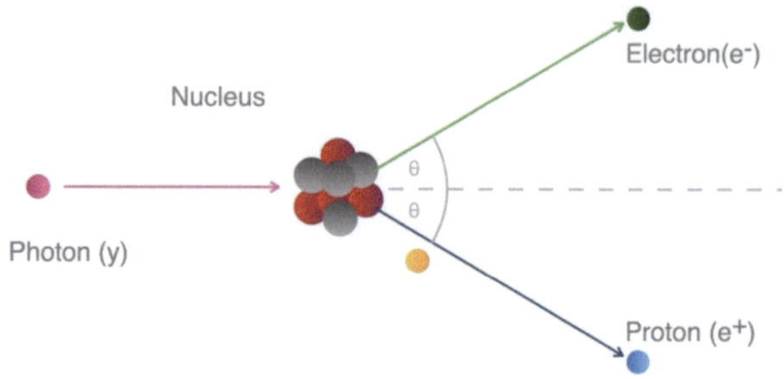

> **Key Concept**
> Blackett and Giovanni PS Occhialini discovered pair production by photographing electrons and positrons created by cosmic rays in a Wilson cloud chamber.

Pair production interactions are useful in positron emission tomography (PET) imaging. These secondary electrons, also called positrons, produced in any of the three types of interactions often have sufficient energy to produce many ionisations that continue until the end of their range.

> **Check Points**
> - What events occur during pair production?
> - What happens to the electrons and positrons that are produced due to pair production effect?
> - What type of gamma rays are responsible for producing pair production?

Thus, alpha and beta rays are composed of charged particles that interact with matter via electrostatic forces. Gamma radiations are electromagnetic waves that interact with medium via the Compton and photoelectric effects.

7 Internal Conversion

Internal conversion is another radioactive decay process in which one of the electrons of the atom interacts with an excited nucleus. This is an alternate process to γ ray emission. Sometimes when a γ ray is released from a nucleus, a collision occurs with an electron in one of the orbits, and it gives up its energy to the atomic electron. The same electron is knocked from the orbit with an energy that is equal to the γ photon energy minus the electron's binding energy in the atom, and this process is known as *internal conversion*. The energy is transferred from an excited state nuclide directly to an orbital electron (present in the K, L, or M shell) instead of to a γ ray. This process is another example of the direct interaction of a nucleus with the orbital electrons revolving around it. Therefore, a high-energy internal conversion electron is released from the radioactive atom instead of a gamma ray. Because of this reason, these high-speed electrons are not called β particles because the latter are released during beta decay and are formed during the process of nuclear decay.

Internal conversion is possible in all cases where gamma decay is possible, except in cases where the atom is fully ionised. During internal conversion, the atomic number remains the same, and therefore (as in the case with gamma decay), no transformation of the element to another takes place. Since this will create a vacancy in one of the inner electron shells, it will automatically be followed by the release of characteristic X-rays, Auger electrons, or both. Thus, high-energy electrons and X-ray photons are emitted, none of which originate from that nucleus. The atom contributed the energy needed to release the electron, which in turn caused the concluding events and other emissions.

Key Concept
Internal conversion is a non-radiative process, as no photons are emitted in the process.

Therefore, a single nucleus will emit either a gamma ray or an electron. However, in a collection of nuclei, some will emit γ rays and some conversion electrons. The number of electrons to γ rays ratio is called the coefficient of internal conversion (IC) for that excited state.

$$\mathrm{IC} = \frac{\text{Total number of electrons emitted}}{\text{Total number of gamma rays emitted}} \tag{2}$$

Check Points
- What is an internal conversion?
- State the difference between X-rays and gamma rays.

Thus, it can be concluded that gamma rays appear through two processes: scattering and absorption. Scattering takes place via Compton scattering, where a gamma ray strikes a free electron present in the medium. A certain amount of the gamma ray energy is transferred to the electron, and the gamma ray continues moving on its path as scattered radiation with a lesser amount of energy but in a separate direction. There are two types of absorption processes during which the gamma ray completely disappears. Photoelectric absorption takes place under conditions in which a gamma ray instantaneously transfers its entire energy to an electron, which causes the release of electrons from the atom. Gamma rays having an energy greater than 1 MeV can also be absorbed by the process of pair production, where a pair of electrons and positrons is spontaneously produced. The positron will ultimately annihilate an electron to produce two gamma rays, each having 511 keV of energy.

The impact produced by gamma rays can be interpreted in the form of linear energy transfer

8 Linear Energy Transfer

When ionising radiation travels through the medium, it gradually loses energy through various interactions along its path. For a particular type of absorber, the rate of energy loss depends on the energy of the radiation, the type of radiation, and the density of the medium. The rate at which electron energy is transferred to a medium is called linear energy transfer (LET) and is usually expressed as the quantity of energy delivered per unit of distance. In tissues, the LET value is inversely related to the electron's kinetic energy. As the electron's energy decreases, its velocity also decreases, and so the LET value will increase until its energy is completely dissipated.

Key Concept
Linear energy transfer determines the biological consequences of radiation exposure.

The SI unit of LET is Newton, but it is mostly expressed as keV/μm or MeV/cm. Because the charged particle slows down as it deposits its energy in the tissue, the LET value varies over the length of the charged particle track. The rate of transfer of energy (−dE/dX, which is the loss of energy per unit distance) increases during this process in such a way that there occurs a peak of energy accumulation at the end of the track (called the *Bragg peak*). The value of LET mainly shows the quality of various categories of radiation and is of importance as its biological effects (i.e., its relative biological effectiveness, RBE) depend on its average LET value. Charged particles normally have a higher LET value than γ rays due to their larger energy deposition along the track. The actual connection between the damage production efficiency and the LET values depends on the biological effect taken into account. For some effects, the efficiency rises with a rise in LET; for others, it falls and for still others it rises until a certain point before falling with a further increase in LET. There is a particular LET value for a specific biological effect that produces the best energy accumulation within the tissue. Also, the radiation is categorised into low LET and high LET radiation. The radiation with low LET values does not produce an appropriate concentration of energy, while the radiation with high LET values has a tendency to deposit more energy than is needed to get the desired effect; this leads to energy wastage and decreases the efficiency of energy transfer. A high LET value will reduce the radiation quickly, thus making shielding effective and also preventing deep penetration. Particulate radiation normally have a high LET value, whereas γ rays have a low LET value because of their scanty ionisations. Generally, the relative biological effectiveness of radiation increases with an increase in LET value up to 100 keV/μm but above this value, it starts declining more than the required energy deposition so as to cause a particular biological effect. Such energy loss events are normally randomly distributed all along the photon track. However, for low LET radiation, the energy deposition events are scanty along the track of the photon with respect to the dimensions of biomolecules such as DNA. Due to this, the photons may pass through such biomolecules without any energy deposition. For such types of radiation, the quantity of energy deposited in an area of the track has similar dimensions as the biological molecules and also varies widely from a few electron volts up to hundreds of electron volts. The energy loss events for high LET radiation are much closer together, and large amounts of energy will be deposited along the path of the track, similar in dimension to biomolecules.

Check Points
- Define linear energy transfer.
- What factors affect LET values?
- What is the difference between radiation with low and high LET values?

Thus, it can be summarised that gamma rays are ionising radiation. The main reaction mechanisms for the interaction of gamma rays with matter are the photoelectric effect, Compton scattering, and electron-positron pair production. Each of these processes occurs in different forms depending on the gamma ray energy. All three of these interactions lead to an attenuation of the gamma ray or X-ray radiation while passing through matter. Internal conversion is another type of decay process during the interaction of gamma rays. It is an atomic decay process in which an excited nucleus strikes one of the electrons in an atom. This results in the release of electrons from the atom. This way it is different from beta decay, where beta particles originate from the nucleus.

9 Characteristic X-Rays and Auger Electrons

Key Concept
The auger effect was discovered by French physicist Pierre- Victor Auger.

Characteristic X-rays were found by Charles Gloves Barkle in 1909, and in 1917, he won the Nobel Prize in Physics for this discovery. X-rays are part of the electromagnetic radiation spectrum. Electromagnetic radiation with energies of approximately 100 eV or more is called X-rays. X-rays are primarily distinguished from other forms of electromagnetic radiation by their ability to produce ionisation in matter and penetrate substances. Characteristic X-rays are produced during the transition of electrons from the outer to the inner orbits of atoms (the K or L shell in most cases). Under normal circumstances, the inner orbits of an atom are fully occupied; therefore, to cause the electronic transition from the outer to the inner orbit, it is necessary to create a vacancy or hole called a "core" hole in the inner orbit. This can be accomplished in various ways. Bombardment of an element with high-energy particles, which can be photons, electrons, or less (such as protons), leads to the production of characteristic X-rays. A well-known example is an X-ray tube, where high-energy electrons sometimes collide with the inner shell electrons, knocking them out of the atoms and thus creating a vacancy in the inner shells. Once a vacancy is created in the inner shell of an atom, an outer shell electron falls to fill this hole. The difference between the potential energies of the two shells involved in the transition is emitted as electromagnetic radiation. This is called a characteristic X-ray of the atom if the energy emitted is approximately 100 eV or more. These X-rays are specific to each element because each element has a distinct set of energy levels, and so the transition from a higher to a lower energy level happens. X-rays with frequencies that are particular to each element are released. If the vacancy was in the K shell, they are known as K X-rays; if the vacancy was in the L shell, they are called L X-rays. Because the energy of a characteristic X-ray emitted by an atom is unique, it is possible to identify an atom (and therefore a particular element) by the energy of its characteristic X-rays. This property is useful in different techniques, like X-ray fluorescence spectroscopy, particle-induced X-ray emission, energy-dispersive X-ray spectroscopy, and wavelength- dispersive X-ray spectroscopy.

An alternate process to characteristic X-ray emission is the emission of *Auger electrons* (named in honour of their discoverer, P. Auger) or Coster-Kronig electrons. Sometimes, instead of dissipating energy as X-rays, the energy is transferred to another electron, which is released from the atom. This is called the auger effect. Auger electrons are the analogue of internal conversion electrons, but here the electron originates from the atom rather than the nucleus. This process (like electron capture) may cause a vacancy in a normally fully filled electron shell. This vacancy is filled by the transition of an electron from one of the outer shells of the atom to the emission of a characteristic X-ray photon. Another option is that the excitation energy may be transferred directly to one of the outer electrons, causing it to be released from the atom. This electron is called an Auger electron and appears with an energy given by the difference between the energy lost by the electron, which made the downward transition, and the binding energy of the shell from which the electron was released. Thus, in the emission of an auger electron, two electrons are removed from the L and/or M shells; one fills the vacancy in the K shell, and the other is emitted from the atom with the rest of the energy. The atom is therefore doubly ionised. A similar process may occur if the vacancy is in the L or M shell.

Key Concept
Auger electrons are extremely low-energy electrons emitted by radionuclides that decay via electron capture (e.g., ^{111}In, ^{67}Ga). This energy is deposited over very short distances (nm or m), resulting in high linear energy transfer with the potential to kill tumours.

The Auger process occurs frequently in elements with low atomic numbers ($Z < 24$, e.g., C, N, O, Al, Ca), whereas X-ray emission is characteristic of elements with high atomic numbers ($Z > 45$, e.g., I, Cs, W, Pb). These two processes, characteristic X-ray emission and Auger electron emission, are diagrammatically depicted in Fig. 4.

The energy spectrum of these electrons does not provide information regarding the atomic energy levels. Auger electrons therefore produce a discrete energy spectrum, with different groups corresponding to different initial and final states. In all cases, their energy is less as compared to beta particles or conversion electrons, mainly because Auger electron emission is preferred only in low-atomic-number elements that have low binding energies of the electrons. Auger electrons with a few keV of initial energy are subject to noticeable self- absorption in the source and are normally stopped by very thin source covers or detector entrance windows. The Auger effect shows resemblance to the internal conversion of the nucleus, which also causes the ejection of an electron. Therefore, in characteristic X-ray emission, energy is released as a single X- ray photon, while in auger electron emission, energy is released by one of the outer electrons leaving, carrying characteristic kinetic energy.

Check Points
- How are characteristic X-rays produced?
- State the mechanism of the formation of auger electrons.

Therefore, it can be concluded that all three major forms of radiation and their interaction with the medium are of clinical significance in diagnostic and therapeutic radiology: the Compton effect, the photoelectric effect, and pair production. Out of these, the Compton effect plays a vital role in modern-day mega voltage radiation therapy. The decreased scattering suffered by high-energy radiation as well as the homogeneous tissue dosage are mainly due to the Compton effect. The photoelectric effect is of primary importance in diagnostic radiology and has only historical importance in present-day radiotherapy. Understanding these interactions may be useful in the study of nuclear detectors, shielding, and the effects of radiation on living organisms.

Further Reading

Albert RE (2001) Ionizing Radiation. Wiley Online Library. https://doi.org/10.1002/0471435139.tox023

Angelis AD, Pimenta M (2015) Introduction to Particle and Astroparticle Physics Multimessenger Astronomy and its Particle Physics Foundations (2nd edn). Springer-Verlag Italia. ISSN 2192-4791. ISSN 2192-4805 (electronic)

Bailey DL, Humm JL (2016) Nuclear Oncology: Physics of Nuclear Oncology. 1–18. Springer

Bennett CJ, Pirim C, Orlando TM (2013) Space-Weathering of Solar System Bodies: A Laboratory Perspective. Chem Rev 113(12), 9086–9150. https://doi.org/10.1021/cr400153k

Bushberg JT, Siebert JA, Leidholdt EM, Boone JM (2012) The Essential Physics of Medical Imaging (3rd edn.) LIPPINCOTT WILLIAMS & WILKINS, a WOLTERS KLUWER business. ISBN 978- 0-7817-8057-5

Fleck I, Titov M, Grupen C, Buvat I (eds) (2021) Handbook of Particle Detection and Imaging (2nd edn.) Springer Nature. ISBN: 978-3-319-93785-4

Hosni HMMA (2010) On some physical properties of GeSe3-Sb2Se3-ZnSe thin films and their radiation response. PhD thesis. NCRRT, Atomic Energy Authority. Cairo University

Knoll GE (2000) Radiation Detection and Measurement (3rd edn.) John Wiley & Sons, Inc. ISBN 0- 471-07338-5 (cloth: alk. paper)

Krane KS (1988) Introductory Nuclear Physics. Oregon State University. John Wiley & Sons Inc. ISBN 0-471 -80553-X

Movahed A, Gnanasegaran G, Buscombe J, Hall M (eds) (2009) Integrating Cardiology for Nuclear Medicine Physicians: A Guide to Nuclear Medicine Physicians. Springer. ISBN: 978-3-540-78674- 0

Podogorsak EB (2006) Radiation Physics for Medical Physicist. Biological and Medical Physics, Biomedical Engineering. Springer Berlin Heidelberg New York. ISSN 1618-7210. ISBN -10 3-540- 25041-7 ISBN-13 978-3-54025041-8

Rambabu N (2017) Interaction of ionising radiation with matter. Kidwai Memorial Institute of Oncology

Tewari KS, DiSaia PJ (2002) Radiation therapy for gynecologic cancer. The Journal of Obstetricsnd Gynaecology Research 28(3): 123-140. https://doi.org/10.1046/j.1341-8076.2002.00036.x

Training Course Series No. 42 (2010) Radiation Biology:A Handbook for teachers and Students. International atomic Energy Agency Vienna. ISSN 1018-5518

van der Meulen NP, Talip Z (2021) Non-conventional radionuclides: the pursuit for perfection. In M. J. Caplan (Ed.), *Reference module in biomedical sciences* (p. (10 pp.). https://doi.org/10.1016/B978-0-12-822960-6.00052-1

Artificial Radioactive Isotopes

Key Learning Objectives

- Primordial or naturally occurring radionuclides
- Cosmogenic radionuclides
- Artificial radionuclides
- Different sources of man-made radionuclides
- Pathways of radionuclides from the environment into the human body

Overview

There is considerable exposure to different types of radiation—natural, cosmogenic, and artificial. Natural radionuclides include primordial radioactive elements and their decay products found in the earth's crust. Cosmogenic radiation is produced by cosmic rays and artificial radionuclides produced by means of a cyclotron, nuclear reactor, or fission process. All these radionuclides reach the human body via the food chain through a number of routes, and many produce several harmful effects in all living beings

As discussed, there are several radionuclides found in the environment, and all these radionuclides are unstable and are continuously disintegrating, releasing high-energy radiation. Different radioisotopes have different and characteristic half-lives. These radioisotopes emit radiation at different rates depending on their half- lives. Radiation has been and continues to be a threat to all living things since their inception. Some of them cause health hazard to humans. Some occur naturally in the environment, while others are man-made, produced either deliberately or as by-products of nuclear reactors. All these are grouped into three main categories:

A. Pathak, *Tools and Techniques in Radiation Biophysics*,
https://doi.org/10.1007/978-981-99-6086-6_6

1. *Primordial radionuclides*— These radionuclides originate from earth long before its creation; these are actually the remnants of the Big Bang cosmogenic sources and ancient supernova explosions that took place before the solar system was formed.
2. *Cosmogenic radionuclides*— These are continuously being produced by cosmic rays hitting the earth; these are produced within the earth's atmosphere, mainly in the stratosphere.
3. *Man-made* or *artificial radionuclides* —These are produced via the process of radio activation using nuclear reactors, particle accelerators, or radionuclide generators.

1 Naturally Occurring or Primordial Radionuclides

The origin of primordial, or naturally occurring, radionuclides is related to the process of nucleosynthesis in stars. Naturally occurring radioactive materials are found throughout nature — in the soil, water, vegetation, and even in our bodies. The main isotopes of concern in this category are uranium and its decay products like thorium, radium, and radon. Some of these are consumed with food and liquids, while others, like radon, are breathed in.

The three isotopes of natural uranium are ^{234}U, ^{235}U, and ^{238}U. Of the three, ^{238}U is the most abundant (99.78%). Thorium is another important primordial radionuclide, about four times more abundant than uranium. Except for radon and thoron and their progeny, these radioisotopes are chemically bonded to minerals in rocks and soils. Thoron and radon are noble radioactive gases that cause lung cancer in greater amounts. According to the WHO, radon and its progeny are the second leading cause of lung cancer after tobacco smoking. If inhaled, thorium may stay in the lungs for a longer period of time. If ingested, it leaves the body via faeces and urine in due course of time. However, a little amount of thorium that is left in the body may enter the blood stream and get deposited in the bones, where it may persist for several years.

Key Concept

The inhabitants of an environment contaminated with radionuclides face detrimental effects of exposure to the radiation, as indicated by an increase in DNA damage and/or oxidative stress commonly associated with radiation exposure.

The radioactive elements that naturally occur are any of four radioactive series that begin with long-living parents and have half-lives of the same age as the Earth.

The following are four naturally occurring series:

- *The thorium series* starts with Thorium-232 (^{232}Th).
- *The actinium series* starts with Uranium-235 (^{235}U).
- *The neptunium series* starts with Neptunium-237 (^{237}Np).
- *The uranium series* starts with Uranium-238 (^{238}U).

The element at which these series begin gives them their names (except the actinium series that begins with a different uranium isotope). All four series start with a parent nucleus that has a very long half-life and eventually degrades into a stable lead nuclide (^{208}Pb) or a stable bismuth-203 (^{203}Bi) isotope through alpha and beta emission. It is presumed that collapsing stars produced all heavy radioactive isotopes in roughly identical proportions; but these elements have different half-lives, resulting in large variations in the quantity of heavy elements nowadays. For example, ^{237}Np has a significantly short half-life as compared to the other parent nuclei mentioned above. It no longer occurs naturally as it has completely disintegrated. The other three parent nuclei (^{232}Th, ^{235}U, and ^{238}U) have longer half-lives and still exist in nature. Low concentrations of uranium, thorium and their decay products are present everywhere. Each series disintegrates via a number of unstable daughter nuclei.

Apart from these, a few naturally occurring radioisotopes lighter than lead (which is the end product of the three naturally occurring decay series) can also be found on the earth. They also have a high isotopic abundance. Most notable among them is ^{40}K, which has a half-life of 1.28×10^9 years. Since it is found in all types of foods, it has the largest proportion of the naturally occurring radiation load via ingestion among humans. Naturally occurring radionuclides are divided into two types. They may occur-

- singly (Table 1); or,
- in the form of series or chains (Table 2).

^{235}U is used to power commercial nuclear power reactors that generate large-scale, low-carbon base-load electricity and also produce isotopes used for medical, industrial, and defence purposes. It can be used as ballast for ships and a

Table 1 Some non-chain primordial radionuclides

S.No.	Isotope	Half-life	S.No.	Isotope	Half-life
1	^{113}Cd	7.7×10^{15}year	**8**	^{144}Nd	2.29×10^{15}year
2	^{142}Ce	$>5 \times 10^{16}$year	**9**	^{190}Pt	6.5×10^{11}year
3	^{152}Gd	1.08×10^{14}year	**10**	^{87}Rb	4.75×10^{10}year
4	^{174}Hf	2×10^{15}year	**11**	^{187}Re	4.3×10^{10}year
5	^{115}In	6×10^{4}year	**12**	^{147}Sm	1.06×10^{11}year
6	^{40}K	1.28×10^{9}year	**13**	^{123}Te	1.24×10^{13}year
7	^{176}Lu	3.73×10^{10}year	**14**	^{50}V	1.7×10^{7}year

Table 2 Primordial radionuclides, which are the components of three series of radioactive elements

S.No.	Nuclide	Half-life
1	^{40}K	1.28×10^{9}year
2	^{226}Ra	1.6×10^{3}year
3	^{222}Rn	3.82 days
4	^{232}Th	1.41×10^{10}year
5	U	7.04×10^{8}year
6	^{238}U	4.5×10^{9}year

counterweight for aircraft. It is also used in ammunition and armour. One of the isotopes of uranium (^{230}U) is also being used to treat diseased conditions using 'targeted alpha therapy', a process where alpha-emitting isotopes are used to treat diseases. Thorium also has nuclear fuel applications, i.e., it is used as fuel for generating nuclear energy. It is helpful in radiometric dating, in manufacturing lenses for cameras and scientific instruments, and in the aerospace industry. ^{225}Ac is used to produce ^{213}Bi, a radioactive isotope used in radioimmunotherapy. It is also sometimes used in cancer treatments and ^{227}Ac is used to study the movement of fluids in oceans, and as an agent for radiation therapy.

Check Points

- What are the three different categories of radionuclides?
- Write a note on primordial radionuclides.
- What are the radionuclides that fall into the category of primordial radionuclides?
- Explain the four series of naturally occurring radionuclides.

Thus, naturally-occurring radionuclides belong to NORM and include radioactive elements such as radium, uranium, thorium, and potassium and their radioactive decay products, as per the Environmental Protection Agency. These have a sufficiently long half-life that allows them to exist in substantially high concentrations in the earth. These three chains of radioactive elements and their decay products and the long-lived primordial nuclide, ^{40}K, account for much of the external background radiation dose from radionuclides to which humans are exposed.

2 Cosmogenic Radionuclides

Key Concept

Cosmogenic nuclide dating uses the interactions between cosmic rays and nuclides in glacially eroded bedrock to provide estimates of the age of rock exposure at the earth's surface.

Cosmic rays are high-energy particles that enter the solar system from outer space. The major source of cosmogenic rays is nuclear interactions due to galactic cosmic rays. There is interaction between these rays and gases and particulate matter present in the earth's atmosphere. These are the rare isotopes formed when a high-energy cosmic ray strikes the nucleus of an *in situ* solar system, thus causing nucleons to be removed from the atom. This type of radionuclide is produced in three stages:

- Formation of primary cosmic rays: Primary cosmic rays (mainly protons) are stable and charged that have been advanced by astrophysical sources situated in the universe.
- Formation of secondary cosmic rays: Secondary cosmic radiation are produced when primary cosmic rays enter the magnetic field of the earth and interact with the nuclei of atmospheric gases. This radiation includes high-energy nucleons and mesons (e.g., muons).
- Formation of nuclides through nucleonic reactions (like spallation, neutron capture, and muon capture): The nuclei interact, causing a "spallation reaction". It is a nuclear reaction in which a highly energetic nucleon strikes a target nucleus and releases protons and neutrons. These accelerated particles further cause secondary interactions and spallation reactions as they interact more with the atmospheric nuclei. The output of this cascade of reactions is the formation of cosmogenic nuclides as well as high-energy radiation (neutrons). Apart from this common method of production of cosmogenic radionuclides, fragmentation, induced fission or capture, neutron capture, and slow muon capture are some other reactions that also contribute to the production of such radionuclides. The production of cosmogenic radionuclides increases with altitude as well as latitude.

These radionuclides diffuse to the surface of the land and the oceans, along with rainfall or particulates associated with them, from where they disperse into ecosystems. Some of the examples of cosmogenic radionuclides are given in Table 3.

Table 3 Cosmogenic radionuclides

S.No.	Cosmogenic Radionuclide	Half- life	Produced by
1	^{7}Be	53.3 days	Cosmic ray spallation of N_2 and O_2
2	^{10}Be	1.5×10^{6}year	Cosmic ray induced spallation of O_2 and N_2
3	^{14}C	5730 year	Low-energy cosmic ray neutron reaction
4	^{36}Cl ^{36}Al	3.01×10^{5} year 7.16×10^{5} year	Cosmic ray interaction with ^{35}Cl, ^{39}K, ^{40}Ca and ^{40}Ar
5	^{3}H	12.4 year	Cosmic ray interaction with ^{14}N and ^{16}O
6	^{81}Kr ^{85}Kr	3.01×10^{5} year 7.16×10^{5} year	Cosmic ray spallation involving protons from ^{80}Kr
7	^{22}Na	2.6 year	By cosmic rays, which induce spallation of atmosphere
8	^{32}P ^{33}P	14.3 day 25.3 day	Spallation reactions induced by neutrons
9	^{32}S	160 year	Neutrons from cosmic radiation bombard ^{40}Ar atoms in the atmosphere

The radionuclides with the highest concentrations in the environment are ^{7}Be, ^{38}Cl, ^{39}Cl, and ^{38}S (Table 3). There are a number of applications related to cosmogenic radionuclides: these are used as chronological markers and universal tracers in dating sediments, continental ice sheets, ground water, the schedule of geologic processes, and radiometric dating (^{81}Kr, i.e., the duration of ground water in old aquifers). The ratios of $^{26}Al/^{10}Be$ in quartz are used to ascertain the age of palaeosols and till layers that are deeply buried. ^{10}Be and ^{26}Al are also used in investigating the erosion rates in a few Antarctic mountain rock samples and to measure the middle Pleistocene erosion rates in buried alluvial sediments. In addition to dating soil organic matter, ^{14}C is utilised to analyse how the atmosphere interacts with the ocean's upper and lower layers. ^{14}C in the atmosphere is also used in carbon dating and ^{32}S is used in groundwater studies.

Cosmic ray protons are continually producing small quantities of radioactive materials. These isotopes are produced in the rocks or soil, in the earth's atmosphere, and also in extraterrestrial items like meteorites.

Check Points

- What are cosmogenic radionuclides?
- Write a note on radionuclide dating.
- Enlist examples of cosmogenic radionuclides.
- What is the source of cosmogenic radionuclides?

Thus, it can be concluded that all living individuals are exposed to background radiation from the air (i.e., cosmic rays), which ultimately leads to the formation of cosmogenic radionuclides. A number of these radionuclides have well-known physical and chemical properties and half-lives. Hence, it is possible to select a suitable nuclide depending on the purpose of the study. Cosmogenic radionuclides are popular for their use as tracers and in dating studies.

3 Man-Made (Artificial) Radionuclides

Key Concept

Irene Curie and Frederic Joliot discovered artificial radioactivity in 1934 by bombarding a sheet of boron and aluminium with alpha particles, which led to the creation of the new radioisotope ^{30}P.

Not all nuclear reactions are spontaneous. This means that all radionuclides are not produced naturally. Instead, a radionuclide can be produced synthetically or artificially by bombarding a substance with some artificial source. These artificial or man-made radionuclides can be produced from the following sources:

3.1 Nuclear Reactors

These devices are used to start and supervise nuclear fission and fusion reactions. The high flow of neutrons activates the elements placed within the nuclear reactor to form radioisotopes. The neutron activation reaction results in the release of gamma rays (Fig. 1). Examples include ^{201}Tl and ^{192}Ir (Table 4).

3.2 Particle Accelerators

> **Key Concept**
> A cyclotron was invented to investigate the nuclear structure by EO Lawrence and MS Livingston in 1934.

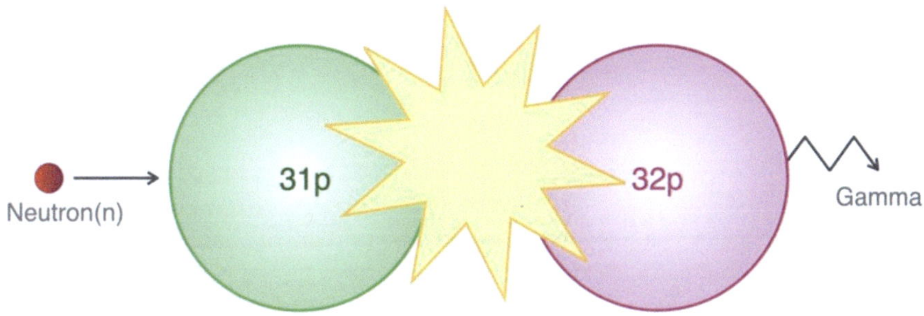

Fig. 1 Neutron activation reaction

Table 4 Research reactor-produced radioisotopes

S.No.	Radioisotope	Half- life	Use
1	^{60}Co	5.27 years	Controlling the cancerous growth of cells
2	^{51}Cr	27.7 days	Label RBCs and quantify gastro-intestinal protein loss
3	^{131}Cs	9.7 days	In the brachytherapy of malignant tumours
4	^{131}I	8 days	Diagnose and treat several diseases associated with the thyroid gland
5	^{192}Ir	73.83 days	Internal radiotherapy is a source for several cancers, like cancers of the head and breast
6	^{99}Mo	66 h	'Parent' in a generator to produce ^{99m}Tc, which is the most widely used radioisotope in nuclear medicine
7	^{32}P	14.3 days	Treatment of excess RBCs
8	^{153}Sm	46.7 h	Reduce the pain associated with bony metastases in primary tumors
9	^{99m}Tc	6.04 h	Image the brain, thyroid, lungs, liver, spleen, kidney, gall bladder, skeleton, blood pool, bone marrow, heart blood pool, salivary and lachrymal glands, and detect infection
10	^{90}Y	64 h	For liver cancer therapy

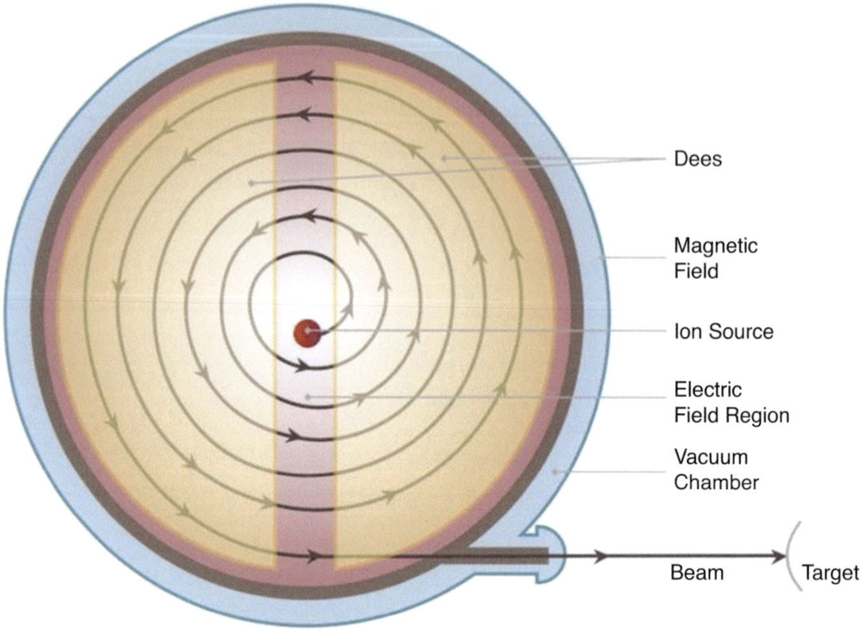

Fig. 2 A cyclotron

A particle accelerator is a machine that produces a stream of charged, high-velocity atomic or subatomic particles and smashes them into a target. A well-known example of particle accelerator is the cyclotron. The cyclotrons are used to accelerate protons so that the protons can bombard a target and produce positrons that release radioisotopes. For example, ^{18}F, ^{15}O, etc. The particles must be fastened to very high energies so as to penetrate the nucleus of the target. A charged particle gains velocity and energy while it is attracted towards an opposite charge. The cyclotron consists of semicircle electrodes ("D"s) within a magnetic field. A vacuum is maintained inside. The "D"s are separated from each other by a thin gap (Fig. 2). High voltage is delivered to the "D"s using a high-frequency oscillator, and a charged particle is placed into the centre to be accelerated. The charged particle will be drawn to the "D" across a "gap". The applied magnetic field ensured that the charged particle moved in a spherical path through the "D", the diameter of which depends on the speed. When the charged particle passes through the first "D", the charge on the "D" has been reversed, and the particle is now attracted to, rather than repelled by, the second "D", gaining speed as it travels across the gap. This procedure is repeated until the particle with the charged charge reaches the target.

There are different types of cyclotrons based on the energy associated with the accelerated charged particle:

- Small cyclotrons (9–11 MeV)—limited to producing ^{18}F for positron emission tomography (PET).

Table 5 Cyclotron-Produced Radioisotopes

S.No.	Radioisotope	Half- life	Use
1	^{60}Co	5.27 years	Controlling the cancerous growth of cells
2	^{51}Cr	27.7 days	Label RBCs and quantify gastro-intestinal protein loss
3	^{131}Cs	9.7 days	In the brachytherapy of malignant tumours
4	^{131}I	8.02 days	Diagnose and treat several diseases associated with the thyroid gland
5	^{192}Ir	73.83 days	Internal radiotherapy is a source for several cancers, like cancers of the head and breast
6	^{99}Mo	66 h	'Parent' in a generator to produce ^{99m}Tc, which is the most widely used radioisotope in nuclear medicine
7	^{32}P	14.28 days	Treatment of excess RBCs
8	^{153}Sm	46.7 h	Reduce the pain associated with bony metastases in primary tumors
9	^{99m}Tc	6.01 h	Image the brain, thyroid, lungs, liver, spleen, kidney, gall bladder, skeleton, blood pool, bone marrow, heart blood pool, salivary and lachrymal glands, and detect infection
10	^{90}Y	64 h	For liver cancer therapy

- Medium sized cyclotron (around 15 MeV)—can produce a large arrangement of PET radionuclides (e.g., ^{13}N, ^{13}C and ^{5}O).
- Large cyclotrons (30 MeV)—are capable of producing other radionuclides such as ^{201}Tl and ^{123}I.

Some of the radioisotopes produced by cyclotron are given in Table 5.

3.3 *Radionuclide Generators*

These devices allow a regular supply of short-lived radionuclides to be accessible on site. Radioisotope generators contain a parent isotope that decays to form a radioisotope. The most common is the $^{99}Mo/^{99}Tc$ generator. ^{99m}Tc is the most widely used radionuclide in nuclear medicine because it has a half-life of only 6.01 h. In this type of generator, a short-lived daughter (i.e., ^{99m}Tc) is paired with long-lived parent (i.e., ^{99}Mo). ^{99}Mo has half-life of 66 h. It is generated by fission and loaded on an aluminium column by adsorption in the generator as ammonium molybdate (Fig. 3).

^{99}Tc and ^{99}Mo have non-identical chemical properties, and therefore, as the parent decays in the ammonium column, the ^{99m}Tc becomes less firmly bound to the ammonium column. Thus, rinsing with normal saline, the effluent, across the column takes off ^{99m}Tc from the column, while ^{99}Mo remains in the column to be decayed to more ^{99m}Tc and is subsequently eluted. The benefit of a radionuclide generator is the availability of carrier- free radionuclides, which improve the stability of radiopharmaceuticals.

Table 6 illustrates the example of some radionuclide generator systems helpful in nuclear medicine.

Fig. 3 $^{99}Mo/^{99m}Tc$ generator

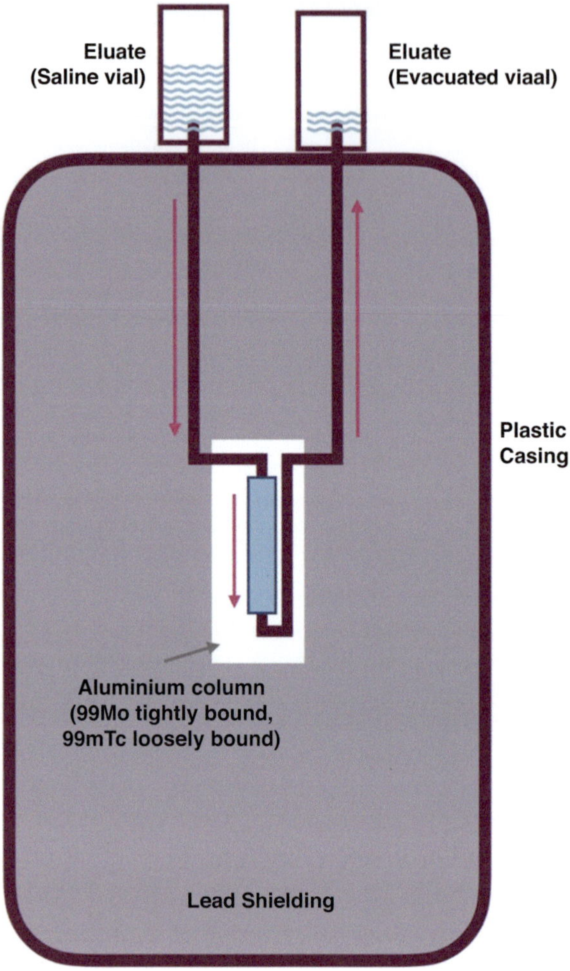

Table 6 Different radionuclide generator systems

S.No	Parent$^{(T_{1/2})}$	Daughter $^{(T_{1/2})}$	Use
1	^{166}Dy (3.4d)	^{166}Ho (1.117 d)	Radionuclide therapy: Treatment of rheumatoid arthritis
2	^{68}Ge (270d)	^{68}Ga (68 min)	Used as a reference source for calibration of 2D and 3D positron emission tomography imaging systems
3	^{99}Mo (2.75d)	^{99m}Tc (6.04 h)	Preparation of ^{99m}Tc- radiopharmaceutical
4	^{81}Rb (4.58 h)	81mKr (13 sec)	Used to evaluate regional lung ventilation
5	^{82}Sr (25.6d)	^{82}Rb (1.25 min)	Used in adults to diagnose heart disease by positron emission tomography
6	^{62}Zn (9.26 h)	^{62}Cu (9.74 min)	Used to produce PET radiopharmaceuticals

3.4 Nuclear Explosions

Nuclear explosions lead to the production of radioisotopes as an unavoidable side effect of nuclear fission and thermonuclear explosions. This process is called fission. Examples of fission nuclides are ^{235}U or ^{239}Pu. Neutron is used to form an unstable nucleus. ^{238}U is converted to unstable ^{236}U when neutron is added to ^{238}U. This rapidly undergoes fission and the fission products rapidly move apart because of high kinetic energy and immediately released as gamma rays and 2–5 neutrons (Fig. 4). These neutrons move fast and are energetic. These are slowed down in a nuclear reactor using scatter reactions and are then used to start subsequent fission reactions thus leading to a chain reaction in which there is uncontrolled generation of energy.

The radionuclides produced by any of the above-discussed methods of inducing a nuclear reaction are called artificial or man-made radionuclides. These types of radionuclides do not exist in nature as such, i.e., there is no natural process that produces them, and they are so unstable that they decay away in a very short period of time (Table 7). These are basically the secondary radionuclides, that are radioisotopes obtained from the disintegration of primordial radionuclides. For example, they can arise in the decay chain of the primordial isotopes ^{232}Th, ^{238}U, ^{231}U, ^{210}Po, and ^{226}Ra. As a result, numerous short-lived daughters in transitory equilibrium with their parents can be traced back to the four radioactive series with long-lived parent nuclei.

Thus, the artificial radionuclides are produced by bombarding stable nuclides with energetic particles derived from nuclear reactors or cyclotrons, or linear accelerators, leading to the formation of new products. The process is called radio activation.

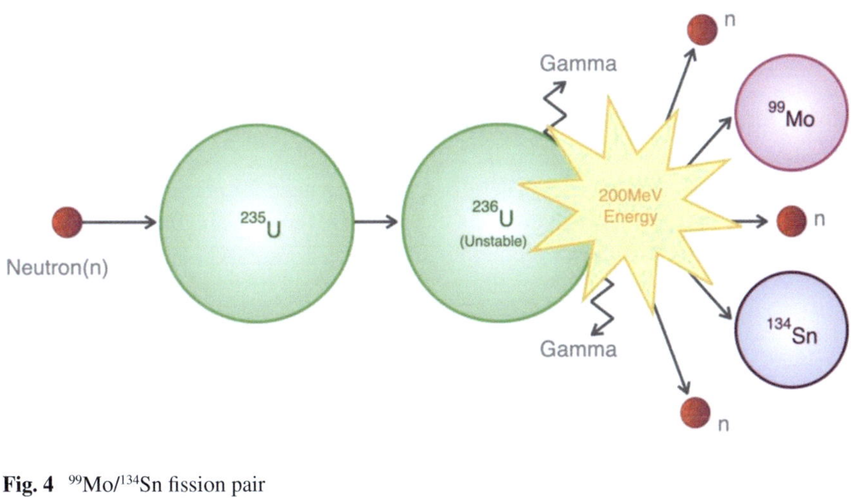

Fig. 4 $^{99}Mo/^{134}Sn$ fission pair

Table 7 Artificially produced radioisotopes and their uses

S.No.	Radioisotope	Half-life	Use
1	^{241}Am	432.7 years	Neutron gauges and smoke detectors
2	^{198}Au	2.7 days	Used to trace factory waste causing ocean pollution and sand movement in river beds and on ocean floors Used to study sewage and liquid waste movements in radiotherapy for cancer treatment
3	^{51}Cr	27.7 days	Studies on red blood cells
4	^{137}Cs	30.07 years	Radio tracing to identify the sources of soil erosion and deposition
5	^{3}H	12.4 years	Tracer to study sewage and liquid wastes
6	^{192}Ir	74 days	Gamma radiography Tracer and study coastal erosion
7	^{54}Mn	312.1 days	To track heavy metals to prevent them from escaping into the environment
8	^{99m}Tc	6.04 h	In diagnostic nuclear medicine procedures
9	^{169}Yb	32 days	Gamma radiography
10	^{65}Zn	244 days	Predict the behavior of heavy metal components in effluents from mining waste water

Check Points

- What are artificial or synthetic radionuclides?
- What is the need for producing artificial radionuclides?
- Explain the different methods of production of artificial radionuclides.
- What is a cyclotron? Explain its workings
- What is meant by "nuclear fission"?
- What are radionuclide generators? Give examples.
- Write the importance of ^{99m}Tc.

It can be summarised that radionuclides are required for diagnostic as well as therapeutic purpose. Naturally occurring radionuclides are unsuitable for these procedures as they have a very long half-life and also they do not have appropriate physical or chemical properties. Therefore, there is a need for producing the desired radionuclides that are short-lived. This can be achieved by using nuclear reactors, cyclotrons, fission, particle accelerators.

These different types of radioisotopes may be present in the environment or may enter into the human body, and the pathways they follow are different in both cases.

4 Pathways of Radionuclides into the Environment

Key Concept
The health hazards from uranium isotopes are due more to chemotoxicity than radiotoxicity.

Every individual is continuously being exposed to radiation from the radionuclides released into the environment because of one of the following activities:

1. *Atmospheric pathway,* i.e., via human activity, radioactive decay, and cosmogenic reactions. Radionuclides can be thrown into the air by human ventures. For example, during nuclear weapon factories, nuclear weapon testing, nuclear power plants, uranium mining, commercial fuel reprocessing, and nuclear accidents. They can also be produced by natural processes, like when cosmic radiation interacts with nitrogen to form radioactive ^{14}C. Natural reservoirs of uranium, potassium, and thorium in the earth's crust also contribute to the natural radiation.
2. *Water pathway,* i.e., deposited in water from the air or from the ground through processes like erosion, seepage, leaching, mining, etc. Radionuclides can come into proximity with water in a number of ways, i.e., they may come from the air when they are released from the atmosphere by rain or snow, released into water from the ground via soil erosion or seepage, or from human activities such as mining or the discharge of radioactive liquids into sanitary sewers, rivers, lakes, streams, or groundwater. Radionuclides also enter ground water from natural deposits in the earth's crust, i.e., from uranium and thorium decay series. They may arise from natural processes in the ground or from human activities. Radionuclides in water can be of concern for human health because they are toxic or carcinogenic.
3. *Food chain pathway,* i.e., radionuclides in water or air, may infiltrate the food chain. Plants, for instance, can take up radionuclides from water. Some of the radionuclides in water will enter the bodies of animals when they consume it. On the surface of plants, radionuclides from the air may potentially condense. The radionuclides are consumed by animals when they consume these plants. Thus, a conduit for radionuclides to get from the environment to people is provided by plants and animals that will eventually become sustenance for humans.

5 Pathways of Radionuclides into the Human Body

Understanding these pathways assists in taking measures to prevent radiation exposure. This can minimise exposure to extra radiation as a result of human activities.

- *Ingestion.* Anything that people eat or drink can have radionuclides. Some are intentionally consumed as part of a medical treatment or diagnostic procedure. While some of these radionuclides are swiftly cleared from the body, frequently within hours, others may stay there for extended periods of time.
- *Inhalation.* This involves breathing radioactive gases, smoke, dust, or particles into the lungs. These radionuclides are suspended in the atmosphere and can enter the lungs. The radioactive gas radon is a natural source of radiation exposure. Radon atoms decays to form other radioactive atoms in the air that are significantly more prone to be retained in the lungs than radon. These "radon-decay products" may be responsible for lung cancer deaths.
- *Through skin.* Radionuclides can also be absorbed via the skin's surface or can enter the body via a break in the skin.

Key Concept
High doses of radiation can induce acute radiation syndrome (ARS) or cutaneous radiation injuries (CRI), as well as cancer in the future.

The dose from terrestrial sources differs in several parts of the world. Areas with higher quantities of uranium and thorium in their soil have greater dose levels. In addition to these sources, all people are exposed to radioactive ^{40}K, ^{14}C, ^{210}Pb, and other isotopes inside their bodies. High levels of radionuclide exposure can damage the cells of the body and can cause cancer.

Check Points

- Explain the different types of human activities that cause contamination of the environment.
- How do radionuclides enter the human body?

Thus, it can be said that different types of radionuclides may enter into air, water, or soil through human activities, through cosmic ray interaction, or by being naturally present in the earth's crust. From here, they may contaminate animals as well as plants and humans through the food chain.

In the end, it can be concluded that the planet on which we live is naturally radioactive. Radiation and radioactivity are essential components of the environment. Every human being is exposed to both natural and man-made radiation. Some are also produced by nuclear reactors and accelerator. Some of the radioisotopes have been a precious gift to the mankind in the fields of medicine and biology. They can also produce harmful and hazardous effects if their levels increase beyond a permissible level in a particular area. Therefore, their optimum use and levels need to be checked.

Further Reading

Arida HA, Ahmed MA (2003) A novel coated graphite rod Th(IV) ion selective electrode based on Thorium oxinate complex and its application. Sensors 3(10): 424-437. https://doi.org/10.3390/s31000424.

Currie GM, Wheat JM, Davidson R, Kiat H (2011) Radionuclide production. The Radiographer: Australian Society of Medical Imaging and Radiation Therapy. 58 (3): 46-52. John Wiley & Sons Australia, Ltd.

Dolenec S, Rebec KM, Lešek A, Šter K, Slovenian LZ, Žibret G, Teran K, Pučko E, Merta I, Poletanovic B, Yaneva E, Kesserű P, Kovács I, et al (2020) Manual for use of Al-containing residues in low-carbon mineral binders. Slovenian National Building and Civil Engineering Institute. ISBN 978-961-7125-00-9 (pdf).

Jadiyappa S (2018) Radioisotope: Applications, effects, and occupational protection. Principles and applications in nuclear engineering. Rahman ROA, Saleh HE-DM (eds). https://doi.org/10.5772/intechopen79161.

O'Shea B (ed) (2010) Textbook of Psychological Medicine (5th edn.). The College of Psychiatry of Ireland

Podgorsak EB (2010) Radiation Physics for Medical Physicists. Part of the book series: Biological and Medical Physics, Biomedical Engineering (BIOMEDICAL) (2nd edn.). Springer. ISBN: 978-3- 642-00875-7

Podgorsak EB (2016) Chapter Radioactivity. Radiation Physics for Medical Physicists: Part of the Biological and Medical Physics, Biomedical Engineering book series (BIOMEDICAL). 262–358. https://doi.org/10.1007/3-540-29471-6_8.

Measurement of Radiation

Key Learning Objectives

- Principle to detect and quantify radiation
- Principle of autoradiography
- Types of gas-filled counters
- Solid scintillation counter
- Liquid scintillation counter
- Semiconductor detector

Overview

Since ionising radiation is invisible and cannot be felt, its detection and measurement must be carried out with the aid of appropriate instruments. These radiation monitoring instruments include a radiosensitive detector and a way to record radiation's effects on the detector. Several physical effects occur during the response of the detector, mainly ionisation. The ion pairs are collected and give an electrical signal that is directly connected with the intensity of radiation. A number of instruments utilise the interaction between radiation and the substances to detect and quantify the high-energy particles.

Radiation measurement is a technique for determining the strength and properties of ionising radiation, like alpha, beta, and gamma rays, or neutrons, with the motive of measurement. Ionising radiation can neither be perceived nor felt. Therefore, it is a challenge to notice its presence. This situation is similar to that of radio and television signals that cannot be seen or felt. But if radio is tuned to the proper frequency, it will find the existence of the radio signal. In the same way,

A. Pathak, *Tools and Techniques in Radiation Biophysics*,
https://doi.org/10.1007/978-981-99-6086-6_7

special detectors or sensors are required to detect ionising radiation. The detector used to detect the radiation must be able to absorb certain parts of the energy conveyed to it. The energy (E_{ab}) will then be used to study some intrinsic processes depending on the type of material required to absorb the radiation, and ultimately, an amount is measured that generates a "signal" corresponding to the initial quantity of the radiation energy (E_{ab}).

Radiation have a wide variety of applications in different fields that include medicine (e.g., PET scan and CT scan), diagnosis, therapy, material analysis, defence, environmental danger monitoring, and security checks in harbours, airports, and borders. In all these applications, radiation detectors play a vital role.

Detection and measurement of ionising radiation are carried out by a number of instruments because these radiations are unseen. Detectors respond to radiation by generating several physical effects that can be gauged. There are four commonly used methods for the detection and quantification of radioactivity. They are established on the following principles:

- The tendency of radioactivity to expose photographic film, i.e., autoradiography;
- Ionisation of gases;
- Excitation of solids and liquids; and
- Formation of electron-hole pairs in semiconductors

The following are the media and related instruments generally used in radiation detection:

- Photographic emulsion (film)
- Gas (examples include the ionisation chamber, proportional counter, and GM counter)
- Scintillator (NaI(Tl), Anthracene, etc.)
- Solid-state detectors (semiconductors, thermoluminescent dosimeters, etc.)

The choice of the detector depends upon the type of radiation, the associated energy, and the intensity of radiation to be measured. Thus, a detailed understanding of various types of detectors and their properties is vital for their selection in a particular situation.

Key Concept
In 1867, Niepse de Saint Victor described the process of autoradiography, which he described as "persistent activity due to an unknown chemical radiation".

1 Methods Based on the Exposure of Photographic Emulsion

Autoradiography is an imaging technique that is used for the localisation of radiation emitted from radionuclides contained within the exposed sample by using photographic emulsions. Thus, it is a photography-dependent technique used to visualise the distribution of a compound that is labelled with either a beta or gamma emitting isotope in a biological sample. Autoradiography generates an image of a radioactive sample by direct exposure of the imaging medium. This technique produces an image of the two-dimensional distribution of a radioactive material. The first autoradiograph was accidentally obtained around 1867 when a blackening was noticed on emulsions containing silver chloride and iodide caused by uranium salts. These studies, along with the research of the Curies in 1898, gave evidence of autoradiography even before the detection of radioactivity. The evolution of autoradiography as a biological process actually commenced after World War II, following the development of photographic emulsions that are made of silver halide.

Autoradiography can provide detailed information on the localisation or binding of radioactive substances in tissues.

1.1 Principle of Autoradiography

Ionising radiation interacts with a photographic emulsion, producing a latent image. For getting a photograph, a source emitting radiation, an object to be detected, and photographic emulsion are required. On a similar line, a radiation source (i.e., radioactivity) releasing from the object and a sensitive emulsion are needed for an autoradiograph, along with a sensitive emulsion. The emulsion contains a large number of silver halide grains present in a solid phase, which is gelatin. Autoradiography is carried out by placing a photographic emulsion along with a substance containing radioactivity in the dark. The detector is actually the silver halide grains of the emulsion. The radioactive material releases energy that goes into the emulsion; the silver halide in the emulsion becomes negatively charged and is then reduced to metallic silver. This metallic silver, during development, appears as black spots, thus forming a particulate latent image. This emulsion is then put in developer and fixer. Photographic developers help to indicate these silver grains as a darkening of the film, and fixatives are used to take off any remaining unbound silver halide (Fig. 1). Thus, a persistent image of the position of the original radioactive event is obtained. Densitometry can be used to estimate the dose of radiation absorbed by a blackened film.

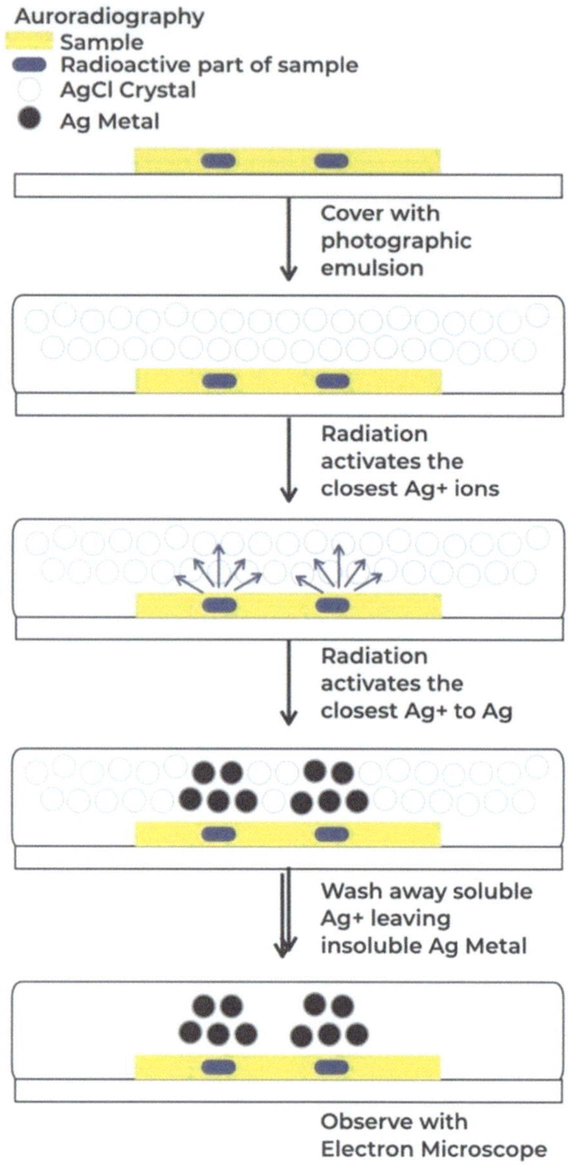

Fig. 1 Events during autoradiography

1.2 *Isotopes Suitable for Autoradiography*

Weak beta emitting isotopes (i.e., ^{3}H, ^{14}C, and ^{35}S) are most acceptable for autoradiography, specifically for locating radioactivity in cells and tissues. The reason for this is that the ionising path of the isotope will be short because of the low energy of the negatrons, and therefore a discrete image will be formed. ^{3}H is the preferred radioisotope for such experiments because all its energy will be quickly released within the emulsion. An electron microscope can then be used to locate the image on the developed film.

For localisation in a whole organism or tissue, either ^{14}C or ^{3}H is suitable, while more energetic isotopes (like ^{32}P) are less acceptable because they release higher energy negatrons that produce much extensive track lengths and therefore less discrete images are formed that are not of good resolution for microscopic studies. On the other hand, for detecting DNA bands in the electrophoretic gel, ^{32}P is preferred because low-energy ^{3}H negatrons, if used, would release their energy in the gel itself and also in the wrap around the gel, thus reducing the sensitivity. However, more energetic ^{32}P negatrons will exit from the gel and form a clear image. If very fine gels are used, then ^{35}S or ^{14}C are useful and can be detected with better resolution; e.g., in DNA sequencing gels, ^{35}S is used as a label. Thus, because of their short path length, low-energy isotopes like ^{3}H, ^{35}S, or ^{14}C are preferred over ^{32}P or ^{125}I because of their better resolution.

Recently, a large variety of biologically significant compounds labelled with low-energy beta emitting radionuclides have become available, and, therefore, rapid growth has been observed in the number of biological investigations employing autoradiographic methodologies.

1.3 *Choice of Emulsion and Film*

An emulsion is an active component in which the image is formed, and it consists of many silver halide crystals suspended in gelatin. The gelatin supports, separates, and protects the crystals. A variety of suitable emulsions are available with varying densities of silver halide crystals packing. A suitable emulsion must be carefully chosen because the resolution obtained depends on the sensitivity of the emulsion.

Following are the three main types of emulsions that can be used for autoradiography:

1.3.1 X-ray Film

This is mainly suitable for large samples like whole-body sections, chromatographs, and electrophoretograms.

1.3.2 Stripping Film

Here, film is available while fixed to a support. It is removed from the support and directly used on the sample.

1.3.3 Liquid Emulsions

These are produced by melting segments of emulsion at approximately 60 °C. Then the emulsion is either poured onto the sample or the sample is dipped in the emulsion. After that, the emulsion is left undisturbed to solidify. This method is called the *dipping film method* and is used when thin films are necessary.

1.4 Autoradiographic Techniques

There are two main methods of autoradiography used in biological investigations. These are:

1.4.1 Temporary Contact Method or Apposition Autoradiography

In this method, the emulsion and the sample are in contact only during the time of exposure, and after the exposure, the emulsion is detached and then developed. The temporary contact method is used mainly for large samples, for example, chromatograms of different labelled materials and leaf and tissue sections with well-defined boundaries.

Very less pretreatment of the sample is required during processing, and any succeeding tissue staining has no effect on the film. Because of imperfect contact, this method generates ordinary resolution and is usually not used for cellular localisation studies.

1.4.2 Permanent Contact Method

There is a permanent contact between the specimen and the emulsion during exposure, while going through photographic processing, and also during histological staining. This method is useful largely with thin samples and is a preferred method for better resolution. This method is used especially for cellular or sub cellular localisation of labelled materials using light and electron microscopes.

Depending on the emulsion being used, there are three modifications to the permanent contact method:

- Mounting method
- Coating method
- Stripping film method

Mounting Method

In this technique, the tissue section is made to float on water, and the emulsion is brought underneath it in such a way that the tissue rests on the film. The film is developed, and the tissue is stained after exposure. This is a basic method with a reasonably high resolution, but it suffers from spotty development because the developer does not penetrate the sample uniformly. To circumvent this issue, the tissue section can be mounted on a slide, and the emulsion can be applied on top.

Coating Method

The emulsion is heated to 60 °C to melt it and is then poured on the tissue sample, where it distributes and solidifies. This method provides higher resolution than mounting.

Stripping Film Method

The stripping films are used and are to be removed from the base and put in contact with the tissue section by water floatation. Good resolution is obtained with this method, which is therefore mainly used for intracellular localisation.

The process of autoradiography can be understood by taking an example: Suppose a radioisotope of suitable specific activity has been given to an organism. To know the tissue where the administered isotope is concentrated, the emulsion film is wrapped around the organism. After a suitable period of time, the film is developed. The blackened region on the film would indicate the position of the organ that has taken up the bulk of the radioactivity, and from knowledge of anatomy, organs can be readily identified. Then, to determine the cell(s) within the organ where the isotope has accumulated, the organ is excised from the animal and enclosed in a photographic film. The cells in the organ that are not taking part in the "trapping" of the isotope would appear as light spots, whereas the cells that take up the isotope would be represented as dark spots. After this, to determine which component in the cell absorbs the isotope, the cells are isolated from the organ, homogenised, and chromatographed. Different components will separate out using the chromatography method after an appropriate time. This chromatogram is then brought in contact with a photographic film, and black spots will show up on the film precisely at the position of the component in which the isotope is present. From the above example, it can be concluded that this technique can be used for a wide variety of studies, starting from gross specimens and going up to the molecular level.

This technique is used with structural parameters in histology and electron micrographs to locate the distribution of a radiolabelled chemical, or it can be applied to chromatographs and electrophoretogram to find out the location of a radiolabelled chemical that can subsequently be recovered and counted with any of the techniques discussed later in this chapter. Even whole-body autoradiographs can be prepared to find out the areas of concentration of administered radiolabelled drugs or chemicals.

Key Concept
Autoradiography is an image on a photographic emulsion generated by the pattern of decay emissions from the distribution of a radioactive substance that is in close contact with the emulsion.

Recent techniques employ a monolayer of silver halide that is applied to ultrathin tissue sections immersed in methacrylate. After emulsion development, tissue sections are stained with heavy metals like uranyl or lead stains and examined under an electron microscope.

1.5 Computerised Image Analysis

Through computer-aided analysis of the optical densities of the autoradiograms, the quantitative determination of the radioactivity concentration in several tissues has been significantly enhanced over the past decade. Image processing systems are either automated microdensitometers or video scanners. Both techniques are inexpensive, quicker, and have the benefit of a variable magnification capacity. The densitometer has a photographic enlarger, a digital multimeter, and the associated electronics. It measured the quantity of light that passes through the film rather than the quantity of light that is transmitted through the air. Some of the advantages of densitometer are the capability to expand the image to accurately measure very "small" areas, a completely unobscured image during measurement, low cost, and ease of use. This type of densitometer is used for quantitative autoradiography. Video-based computerised image processing systems contain a video camera, a microcomputer, graphic components, monitors, picture memory, a digitising pad, etc. by subdividing the video into pixels. The higher the pixel density, the sharper the image that is formed on the monitor. The computerised programmes allow control of contrast, pseudocoloring, etc. In addition, it permits interactive annotation and analysis of the autoradiogram, including extremely minute and irregular distributions. In some programmes, even images stored in memory can be rotated and translated, and two images can be superimposed for comparison. The superimposed pictures may be manipulated pixel by pixel. Various software programmes for image improvement have also evolved. Earlier systems for image exploration were very costly, mainly the computer hardware, and not many programmes were available for several image refining applications, rendering them hard to handle for the examination of the autoradiograms. Therefore, the technique of autoradiography has been markedly improved by computerised exploration of the image and processing of the autoradiograms. Though alternative methods are now available, one can anticipate that autoradiography in the years to come will also provide an important mechanism in various fields of research.

Key Concept
In 1924, the first biological experiment using autoradiography mapped the diffusion of polonium in biological specimens.

1.6 Practical Problems Encountered During Autoradiography

The following things should be kept in mind while performing autoradiography:

(a) **Background**: Accidental light exposure, compounds in the sample, natural background radioactivity (especially ^{40}K in glass), and even pressure exerted during film handling and storage can result in a background fog (or latent image) on the film. This can be difficult, especially in high- design work (such as microscopy), and caution must always be taken to neutralise its effect.

(b) **Thickness of section and photographic emulsion**: The thickness of section and of photosensitive emulsion is another factor affecting the resolution. The thicker the emulsion, more shall be a deposition of silver, thus making the image disproportionate. These parameters are effective with high-energy emitters but not that much with tritium. Better resolution with high- energy emitters can be achieved by the use of thin sections of the material and a thin photosensitive emulsion. The overall adjusted situation is to use thin sections of tissue with thin emulsion layer, and the two are placed in close contact. This will give optimum resolution to observe the results with the utmost precision. The properties of emulsion and the type of developer also affect the resolving power.

(c) **Time of exposure and film processing:** Background will increase with an increase in exposure time, which should be maintained to a minimum every time. The type of isotope, the type of sample, the intensity of activity, the type of emulsion film, and the aim of the experiment all determine the time of exposure.

During film processing, the chemicals are used to develop the exposed emulsion, which contains a reducing agent that functions better in a slightly alkaline medium. The reducing agent converts the silver bromide to metallic silver, and metallic which is deposited to form enlarged grains. The grains that have been developed by the reducing agent are fixed to the film to finally come out as a negative image. The common developers are hydroquinone, amidol, and metol. The fixers used during processing are also acidic. Fixative contains 25– 30% hypo (sodium thiosulphate) that makes complex with deposited silver and dissolves the non-reacted or less reacted silver bromide proportionately, and hence hues and tones are developed. With the progression of fixation, the opaque and latent image becomes more and more transparent. Usually autoradiographs are fixed over twice the duration required for emulsion to clear, but this duration should not be allowed to exceed as that can dissolve out the deposited silver grains as well. After completion of developing and fixation, the film can be viewed for the sections in which the pattern of silver grains is well differentiable.

Thus, the quality of the image depends on a number of factors, and with proper care, the background can be kept within acceptable limits.

1.7 Disadvantages

There are a number of difficulties encountered during autoradiography. Some of these are listed below:

- Autoradiography is a very sensitive technique. However, the sensitivity largely depends on the type of radionuclide and the purpose of investigation; e.g., if radioactive receptor ligands are used, the density of receptors can be calculated; the amount of RNA can be measured if *in situ* hybridisation is carried out; and the number of adducts can be obtained if covalent linkage of bioactivation products is studied.
- In many instances, structures are examined that are so minute that it is nearly impossible to differentiate them for other types of analyses. This may be the case, for instance, in studies concerning the localisation of pharmaceuticals during embryogenesis.
- The major disadvantage of *in vitro* autoradiography is the fact that radioactivity is involved, and detection depends on the rate of decay of the radioisotopes. Therefore, there must be an exposure period that makes it difficult to obtain results rapidly.
- Moreover, when using long-lived radioisotopes, image development may need longer periods of exposure, making it very time-consuming.
- Also, there is a lack of standard protocols to find the binding specificity of radiotracers to their targets. The non-specificity of the ligands can easily lead to misinterpretation of the results.
- Also, all the autoradiography methods use ionising radiation to produce the final image. In PET and SPECT scans, patients are exposed to a higher dose of radiation due to internal exposure to the radioactive source. So there is an increased risk of health conditions such as cancer and neurodegenerative diseases during such scans.
- Autoradiographic results are mostly unsatisfactory in relation to quantification and reproducibility. This is partially due to the nature of the methods and cannot be eliminated entirely. For instance, it is impossible to restore the initial distribution of silver bromide grains after development.
- Insufficient consideration of the photographic fundamental processes and the characteristics of the emulsions employed frequently leads to erroneous interpretations of autoradiographic results.

- Artefacts in the form of positive and negative images, caused by chemical processes in the specimen or image fading, particularly during extended exposure time or improper storage environments, are an additional issue.
- There is also a physical distortion of the film, such as at the specimen's edges or due to inappropriate handling.
- If the radioactive label vanishes from the primary molecule and gets excreted, or if it takes part in the intermediary metabolism, *in vivo* drug distribution studies may produce misleading results. Tissue extract chemical analysis is therefore frequently advised.

However, in the case of whole-body (or macro) autoradiography, abundant information is obtained that no alternative technique could provide.

1.8 Applications

There are several radioisotopes that can be used in autoradiography (Table 1).

This process of autoradiography has high sensitivity and has been useful in several biological experiments. The autoradiographic technique is considered a development of histochemistry. These require the location of the radioactivity in different types of biological specimens. For example, the location of the sites of radiolabelled drug sites in the body of an experimental model can be found by placing the model adjacent to a sensitive emulsion. After a specific period of exposure and following the development, the plate will display an image of the section in tissues and organs with high radioactivity.

The autoradiographic technique offers a number of applications, as listed below:

(a) *Macroscopic studies*: Radioactive metabolites that have been isolated and separated by chromatographic or electrophoretic methods can be identified on the chromatograph or electrophoretogram, and the radioactive areas can then be recovered for counting and identification (counting may be performed on the

Table 1 List of some radioisotopes along with their uses

S. No.	Parameters	Ionisation chamber
1	^{40}Ca	Studying cellular functions and bone formation
2	^{14}C	To study metabolic reactions
3	^{137}Cs, ^{67}Cu	Used in cancer treatment
4	^{51}Cr	RBC studies
5	^{57}Co	Pernicious anaemia
6	^{131}I	Treat thyroid disorders
7	^{75}Se	Protein studies
8	^{3}H	Drug metabolism
9	^{133}Xe	Lung ventilation and blood flow

original chromatograph using a chromatograph scanner, the design of which is generally based on GM tubes, or by elution from the paper, plate, or gel for counting by liquid scintillation counting, as will be discussed later in this chapter).

(b) *Localisation of cells/tissues*: In cell biology studies, this technique may be useful to determine the localisation of a radioactive material in a tissue or cell that is incorporated into a metabolic pathway, associated with a receptor or enzyme, or hybridised with a nucleic acid. In combination with microscopy, autoradiographic techniques are used to localise the label in tissues and cells and also for comparative analysis of the incorporated radioactivity.

(c) *Hormone/receptor studies*: In the field of behavioural endocrinology, autoradiography is useful in determining hormonal uptake and specifying the location of receptor. The study can be carried out *in vitro* or *in vivo*. For *in vitro* studies, an animal can be administered with a radiolabelled hormone. *In vivo* autoradiography involves the use of radiolabelled ligands to map out the locations of receptors in living tissue. The ligands are normally labelled with ^{3}H or ^{125}I. Autoradiography is used to ascertain receptor distribution and localisation in the study of neurodegenerative disorders.

(d) *Cell cycle events*: It is possible to introduce radioactive precursors of DNA and RNA, [^{3}H]- thymidine and [^{3}H]-uridine, to living cells in order to determine the time line of several phases of cell cycle. The sequences of RNA and DNA virus can also be identified in this manner. Typically, these instruments are labelled with ^{32}P, ^{33}P, or ^{35}S.

(e) ^{85}Kr is utilised to examine aircraft components for minute flaws. ^{85}Kr is permitted to permeate minute fissures, and its presence is then detected using autoradiography. The technique is referred to as "*krypton gas penetrant imaging*." The gas can penetrate smaller apertures than liquids used for dye penetrant and fluorescent penetrant inspection.

(f) *Kinetic studies:* In metabolism, autoradiography provides information on the distribution and elimination kinetics of a compound in various organs.

(g) Drug distribution: Autoradiography has been used for quantification and localisation of drugs in tissues and cells for decades. This provides pharmacokinetic compartmental analysis, which has been useful in toxicology, pharmacology, drug disposition and patterns, and to predict human exposure to drugs and metabolites, and as well as radioactivity. Autoradiography is also used to quantitatively resolve the localisation of radiolabelled compounds at the cellular level in a histological preparation.

(h) Incorporation of precursors in macromolecules: As a DNA precursor, labelled thymidine can be used to study DNA synthesise and cell proliferation; uridine can be used for the study of RNA synthesis; and amino acids can be used for the study of protein metabolism. Labelled hormone precursors and hormones are used in order to study sites of hormone formation and to localise hormone receptors.

(i) *In situ* hybridisation: *In situ* hybridisation histochemistry refers to the identification of RNA transcripts in tissue sections using radiolabelled, complementary oligonucleotides or ribonucleic acids ("riboprobes"). The objective of quantitative autoradiography (QAR) *in situ* hybridisation is to ascertain the amount of radioactive oligonucleotide or riboprobe found in the area of tissue silica that generated the autoradiographic image. For example, if a radioisotope of iodine (^{131}I) is injected into an animal, it accumulates in the thyroid cells. It can be localised by making a histological section of the thyroid gland and covering it with a photographic emulsion. Subsequent development, fixation, and staining show localised blackening of the emulsion due to radiation emitted by the ^{131}I, and it revealed the concentration of ^{131}I in the colloid of thyroid vesicles. Similarly, ^{32}P has been used for the studies on the demonstration of the movement of phosphate in the transpiration stream of plants and phosphate turnover in bone, etc. But the difficulty arises in obtaining good resolution with the technique because the radiations are not strictly confined to the point from which they are emitted. For these reasons, ultra-thin sections and weak radiating isotopes like ^{3}H are being used. When a ^{3}H-labelled amino acid is injected into an animal, it is rapidly incorporated into cell's proteins, and the location of this protein within the cell can be identified. The thinness of the sections allows examination by electron microscopy.

(j) Glucose utilisation: ^{14}C- 2-deoxyglucose and ^{14}C-iodoantipyrine are used for studying glucose utilisation in the nervous system and in extra-neuronal tissues in both humans and animals. This technique has led to the characterisation of regional cerebral glucose metabolism in animals under a variety of experimental conditions. This method has widespread application as a technique for determining neuronal responses to physiological and pathophysiological actions, with the added advantage that the entire brain can be examined simultaneously in autoradiographic sections with considerable histological detail.

(k) *In vitro* receptor autoradiography: Several hormone and neurotransmitter receptors are visualised by incubating tissue sections with labelled ligands for their respective receptor sites. ^{3}H or ^{125}I are generally used to label the ligands.

(l) Blood flow studies: For these studies, ^{14}C-iodoantipyrine is mostly used. Autoradiography can be used to measure the regional blood flow of the renal cortex, spleen, lung, skeletal muscle, bone marrow, thymus, testes, and other organs.

(m) Localisation of bioactivated products of xenobiotics: For foreign substances, it can be used to study the accumulation in target sites. For example, compounds that are accumulated and retained in foetal organs can give an indication of sites of teratological effects. Chemical carcinogens and other toxic compounds may undergo metabolic activation in the tissues, and autoradiographic techniques can be used to localise the sites of the bioactivation.

(n) Autoradiography in the clinical context: Advanced autoradiographic techniques like positron emission tomography (PET) and single-photon emission computerised tomography (SPECT) are utilised in the clinical setting for the diagnosis and monitoring of diseases. They differ from classic autoradiography in that they use different radiotracers that are appropriate for use by humans, and decay is monitored by a gamma camera that produces high-resolution three- dimensional images (called tomograms). PET is a diagnostic imaging technique used to observe biological processes in the body. Here, a radiotracer is administered into the body. The decay of radiotracers produces positrons and ultimately leads to the production of two gamma rays, which are detected by a gamma camera. SPECT and PET scans are similar, but they use different radiotracers. Here, radiotracers directly emit gamma rays. These two techniques will be discussed in detail in the succeeding chapters. Utilising radionuclides with brief half-lives, such as C ($Tf = 20$ min) and ($Tf = 110$ min), in autoradiography has recently gained practical interest as a compatible technique to positron emission tomography (PET). Since the resolution of autoradiography is approximately 50 times greater than that of PET, the results of autoradiography investigations may assist in the interpretation of PET images. As PET technology continues to develop, C- and ^{18}F-labelled substances will increase, allowing for a significant decrease in exposure durations and double-isotope studies, as will be discussed shortly. In principle, autoradiography with short-lived radionuclides is analogous to conventional autoradiography, but the various steps must be accelerated. Higher radio concentrations (approximately 50 pCi/g body weight) and a short observation period of the experimental animals after administration of the labelled compounds are required (less than 3 half-lives of the radioisotope). With regard to the planning of, say, current PET investigations, the short exposure period is obviously an advantage. It is also possible to achieve high specific activity in the range of 200–2000 Ci/mmol. This is especially important when working with highly toxic compounds and substances that have potent physiological effects, such as peptide hormones.

(o) *Double-tracer autoradiography*: Double-tracer techniques help in simultaneous mapping the dispersal of two radiolabelled substances in the same specimen. The use of one short-lived and another long- lived radioisotope, such as C and ^{14}C, has proven to be the most promising method in this regard. There are two distinct exposures. The first, being very short and providing the distribution image solely due to the short-lived nuclide, is negligible at this point in time. The subsequent exposure, which occurs when the short-lived isotope is almost completely disintegrated, is relatively prolonged and yields an autoradiogram of the long-lived isotope only. This method uses ^{131}I-iodo antipyrine and ^{14}C-deoxyglucose, respectively, to measure cerebral blood flow and cerebral metabolism simultaneously.

1.9 Refinements in Autoradiography

Electron microscope autoradiography reveals information on the ultrastructural localisation of radioisotopes in biological tissue and involves *in vivo* labelling, processing for microscopy, and qualitative and quantitative analysis of the final preparation. The technique's resolution and sensitivity depend on the energy of radioisotope emission and numerous autoradiographic preparation factors. Transmission electron microscope (TEM) or scanning electron microscope (SEM) can be used to perform high-resolution autoradiography. Scanning electron microscope autoradiography is distinguished from transmission electron microscope autoradiography by the bulkiness of the specimen in the former case. In the case of SEM autoradiography, the silver grains generated in the emulsion may originate from either the internal areas of the specimen or the radioactivity of the cell surface. Unlike SEM autoradiography, TEM typically uses ultrathin specimens that have a higher spatial resolution. However, the small quantity of label present may necessitate lengthy exposure times for an adequate yield of results. This type of autoradiography has been utilised with a variety of labelled compounds and isotopes, such as ^{3}H, ^{45}Ca, ^{125}I, and ^{32}P.

It appears that future development in electron microscope autoradiography will involve more diverse and newer applications of the technique, as well as more advanced computer analysis of autoradiographs, rather than significant resolution or sensitivity improvements.

1.10 Radioactive Labelling of Cells or Tissues

The radioactive isotope may exist as an isolated molecule, such as ^{45}Ca cations, or as part of a complex molecule, where the nuclide (such as ^{125}I) is bonded. In the second situation, the radioactive element may function as a specific probe for certain receptor molecules or as a specific precursor for the creation of a particular macromolecule (such as thymidine for DNA synthesis). In some instances, the labelling procedure may also involve the successive use of non-radioactive precursors (for example, in a pulse-chase experiment), specifically when intracellular synthesis and transport routes are being investigated. After labelling, specimen preparation entails fixation, embedding, and sectioning, followed by coating with photographic emulsion and incubation in the dark for a predetermined period of time. Before microscopic observation, the preparation is ultimately developed photographically and fixed. The location and quantity of silver granules on the specimen can be evaluated qualitatively through direct observation and quantitatively through statistical analysis.

Other cytochemical techniques, such as immunohistochemistry and X-ray microanalysis, present a rather static view of the cell. However, the ability to label cells to track the time course of chemical reactions within the protoplasm is a significant advantage over these methods. The success of autoradiography is highly dependent on the extent of labelling and its specificity. Future use of this method will coincide with the creation of extremely targeted site-specific probes (labelled molecules).

Check Points

- State the different methods of detecting radioactivity.
- Write down the principles of autoradiography.
- Which isotopes are preferred for autoradiography?
- Explain the different types of emulsion films used in autoradiography.
- Describe the permanent contact method.
- Explain the factors affecting autoradiography.
- Write the applications of autoradiography.

Therefore, autoradiography can be regarded as a specific biological tool that is used to locate radioactive substances by using X-ray photographic films. It is a straightforward and sensitive photochemical technique used to record the spatial diffusion of radiolabelled molecules within a specimen. These days, autoradiography has been widely employed as an important detection tool that helps in a better understanding of various molecular pharmacological pathways. Recently, autoradiography has been extensively used to study many medical and biological problems. However, this technique requires a sensitive medium, photographic emulsion, and therefore antiquities in autoradiographic films may occur easily and may lead to erroneous conclusions. Therefore, there is a need to use all conditions wisely, and preventive measures should be taken so that the background is reduced and a clear image is obtained.

2 Methods Based on the Ionisation of Gases

Gas-filled radiation detectors are the most classical of all radiation detectors that are based on the ionisation of gases and are still being used in various applications of radiation protection surveillance programmes. These detectors are popular because they are inexpensive, simple to operate, available in various shapes and sizes, and are reliable.

A gas-filled detector consists of a chamber loaded with a gas (therefore named gas-filled detector) and an electronic circuit (Fig. 2). The chamber contains a positively charged metal tube (the cathode) that is filled with gas. An inert gas, for example, argon, xenon, or helium at low pressure, is used. An inert gas is used to prevent chemical reactions from occurring within the gas after ionisation, that could alter the properties of the detector. Argon is mainly used as a mixture of 90% argon with 10% methane (called P-10 gas) due to its low cost. For finer gamma ray detection, krypton or xenon gas is used in place of argon. The type of gas to be used depends on the purpose of the counter. The migration of free electrons takes place during gas multiplication. That is why the gas must not absorb electrons. Air (specifically O_2) is not preferred due to the formation of negative oxygen ions. On one

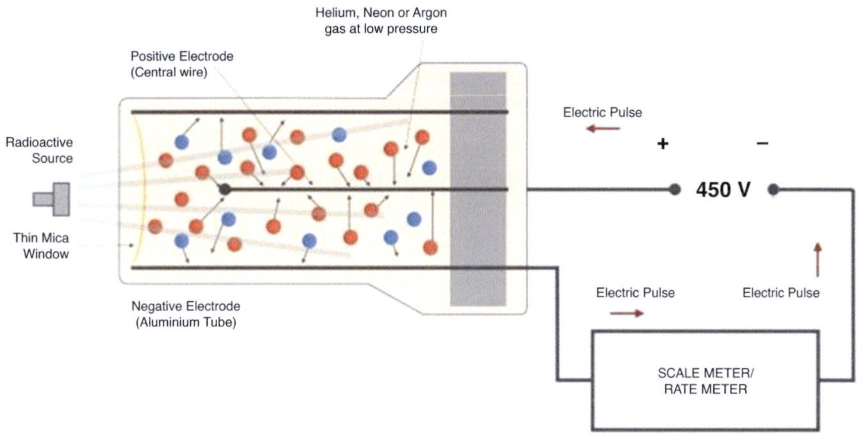

Fig. 2 Gas-filled detector

side of the cathode chamber is a thin mica window through which the radiation can enter the chamber. The thickness of window or wall material is of the order of a few microns or a few millimeters. An anode is a thin metal filament located near the chamber's centre. Low-energy rays and charged particles cannot penetrate the counter, but high-energy rays and neutrons can easily pass through. An external power source is connected to the anode and cathode. The cathode and anode are charged with negative and positive charges, respectively, as their dimensions and the voltage of the power source allow.

In the absence of radiation, the gas between the anode and cathode functions as an insulator, and the electrical charge does not move. As the radiation (γ rays, X-rays, or charged particles) with energy more than 10 eV enters the detector and moves through a gas, the electrostatic field dislodges orbital electrons from elements sufficiently close to its path, causing ionisation. The amount of energy needed to bring about ionisation is dependent on the nature of the gas present in the chamber but is normally between 20 and 45 eV per ion pair. Ions are formed through a secondary process in which some of the particle's energy is transferred to energetic electrons. These free electrons and positive ions move towards the opposite electrodes, respectively. The number of ion pairs produced is collected and measured as current. The current exhibited increases as more radiation enters the chamber. The number of electrons measured depends on a number of factors, including the number of charged particles and/or photons being measured (i.e., the strength of the radiation source or field), the energy of the radiation, the geometric configuration of the detector, the composition of the gas in the chamber, the volume, pressure, and temperature of the gas, and the applied voltage between the anode and cathode. The most important parameter, however, is the voltage applied, V, between the two electrodes. The electrical potential difference between the anode and cathode drives electron movement, and its magnitude shows how many electrons reach the anode and what ensues to them en route.

The ability to induce ionisation the of gas molecules decreases in the order

$$\beta > \alpha > \gamma\left(10{,}000:100:1\right) \tag{1}$$

This shows that α and α particles can be detected by gas ionisation methods, but these methods cannot detect γ radiation and soft β particles.

2.1 Voltage Response Curve

The voltage applied between the anode and cathode determines the number of electrons produced and hence measured. If this voltage is varied, the number of ion pairs collected can be described in the form of six different regions of response (Fig. 3) in the curve.

The curve has six regions:

1. Recombination region
2. Ionisation region
3. Region of proportionality
4. Region of limited proportionality
5. Geiger-Muller region, and
6. Region of continuous discharge

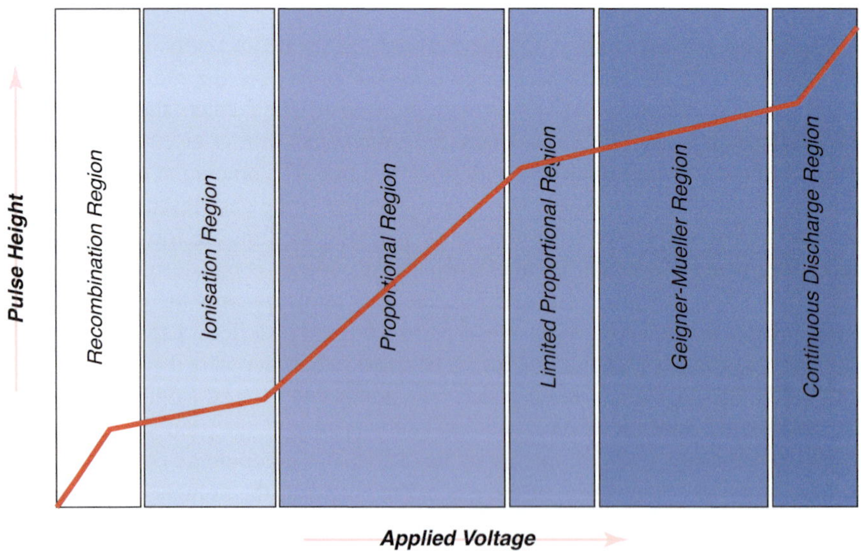

Fig. 3 Voltage response curve for a gas-filled detector

2.1.1 Recombination Region

In the recombination region, the voltage is low such that ion pairs generated by a ray or a particle (called primary ion pairs) are moving so slowly that they can easily recombine to form neutral atoms or molecules, i.e., the electrodes only capture some of the primary ion pairs. As a result, the voltage pulse height is relatively low. With an increasing applied electric field, the mobility of the ions increases, so the time for recombination is less, and the amount of charge that is collected by the electrode becomes larger. As the voltage is increased, more number of primary ion pairs are gathered by the electrodes, and the detected signal increases. There are no viable detectors that work in this region of the voltage response curve.

2.1.2 Ionisation Region

In the ionisation region of the curve, the voltage is high enough so that all primary ion pairs produced are attracted towards the electrodes, and there is no recombination, i.e., due to the high electric field, the ion pairs are forced to move towards the electrodes, and because recombination is delayed or stopped, more ion pairs reach the electrodes. The amount of current produced does not alter significantly following an increase in the voltage because all primary ion pairs produced by the radiation have been collected. The voltage sufficient for this saturation point to be attained is the *saturation voltage*. There is a broad voltage range in which the response is uniform, so even if the voltage varies, a similar number of electrons will be gathered. The size of the signal is therefore proportional to the rate of interaction between ionising particles in the detector. The detectors working in this region are called ionisation chambers. For example, a dose calibrator and an ionisation survey meter. The quantity of current produced by a single ray or particle (total number of primary ion pairs) is still insignificant and cannot be detected unless the interaction of many rays or particles is taken into account. Therefore, due to the low currents, highly sensitive measuring instruments are required. This region is rarely used in quantitative research. However, due to the small size of the produced signals, they are typically reserved for use with large fluxes of radiation or heavy charged particles.

2.1.3 Region of Proportionality

At a voltage level greater than that required in simple ionisation chambers, electrons resulting from ionisation move towards the anode much more rapidly; consequently, they cause secondary ionisation of gas, thereby resulting in the formation of secondary ionisation electrons that cause further ionisation, and so on. Thus, an increase in the voltage generates a signal greater than but proportional to the number of ion pairs produced by the radiation. Thus, from the original event, a whole torrent of electrons reached the anode. This principle of gas amplification is known as the Townsend avalanche effect, named after its discoverer. Because of this gas amplification, current flow is much greater, thus leading to the multiplication of charge.

Gas-filled detectors that work in this region are called *proportional counters*. The total number of ion pairs collected is proportional to the voltage applied (in a range of 1000–2000 V) until a particular voltage is reached, at which point a plateau is formed. Ion-multiplication up to 10^3 to 10^5 can be attained by this method of operation. These are mainly used for detecting and distinguishing between α and β particles and to identify radionuclides on the basis of their decay products. Proportional counters are useful in a number of applications of physics but are not frequently used in nuclear medicine. The pulse depends on the type of radiation, but more crucially, it depends on the incident energy of the radiation. The primary disadvantage of counters designed to operate in the proportional region is that they require a very stable supply of voltage. Therefore, minor voltage fluctuations result in significant changes in amplification. Proportional counters are notably useful for the measurement and quantification of alpha emitting isotopes; however, relatively few of these isotopes are used in biological research.

2.1.4 Region of Limited Proportionality

The proportionality between charge collection and energy deposited in the detector terminates at a certain value, depending on the detector's characteristics (such as size or gas). There is a region known as limited proportionality before the plateau is attained (upper end of the region). Positive ions move slowly and take a long time, whereas free electrons are rapidly collected due to their high mobility. A cloud of positively charged ions is produced during the collection of electrons, and it disperses slowly. If the concentration of positive ions is more, the electric field gets distorted, leading to disruptions in gas multiplication and other nonlinearities. Therefore, the increase in the applied voltage results in non-linear effects. Here, the emission is more voltage-dependent than the initial ionisation.

This is also often not used in the detection and quantification of radioactivity.

2.1.5 Geiger-Muller Region

In the Geiger-Muller region, all particles, including weak beta radiation, cause complete ionisation of the gas in the chamber. The voltage is so much higher that the primary ion pairs produced by a ray or particle acquire sufficient energy to produce a greater number of secondary and tertiary ion pairs, which then collide with the metallic electrodes and additional neutral gas molecules to produce additional ionisations and excitations. In addition, these electrons excite other gas molecules also through collision interactions, which they de-excite by emitting ultraviolet (UV) radiation. These photons interact with other gas molecules by photoelectric interactions, resulting in the production of more free electrons, which are then accelerated towards the anode, causing additional ionisations and excitations. The de-excitation of the molecules produces ultraviolet light, which increases ionisation. Consequently, each radiation event generates an avalanche of ions in the chamber. The amplification factor may reach 10^8, and the detectors operating in this region are extremely sensitive. As a consequence, the total volume of gas is in a temporary state of discharge, and a significant amount of current is generated in response to each radiation event, which is largely independent of the voltage and the energy of the radiation. The pulse size remains the same for all radiation events and becomes independent of the type of radiation or the quantity of energy transfer.

Geiger-Muller pulses contain neither energy nor particle information, i.e., there is no difference in pulse height for particles of various energies and types. Because maximum gas amplification occurs in this area, the magnitude of the detector's output pulse will be constant over a wide voltage range (i.e., the Geiger-Muller plateau). This area, also known as the Geiger-Muller plateau, is where the Geiger counter, a widely used radiation detector, operates. As a result, the magnitude of the current is independent of the number of primary ions produced. This pulse's frequency is measured, not its magnitude. Consequently, it is impossible to distinguish between different types of isotopes by using this form of counter.

2.1.6 Region of Continuous Discharge

In a continuous discharge region, radiation is not required to generate discharge because the voltage is so high. The strong electric field ejects the electrons from atomic shells; atoms and molecules become ionised, and a discharge can occur even in the absence of radiation. There is total ionisation of the gas between the electrodes.

No useful detector can be operated in this region because operation of a detector in the continuous discharge region can cause permanent injury to the detector. There are no useful radiation detectors in this area. Due to the fact that it takes a specific amount of time for the ion pairs to reach their respective electrodes, so other ionising particles that enter the tube during this time do not produce ionisation and are therefore not detected, thereby reducing the counting efficacy. This is known as the tube's inactive time or dead time and is typically between 100 and 200 μs. The ions are neutralised when they reach the electrode. Eventually, some will escape and create their own avalanche of ionisation. A GM tube would therefore be inclined to discharge continuously if left unchecked. To counteract this, the tube is quenched by adding suitable chemicals that reduce the ion's energy. Common quenching agents include ethanol, ethyl formate, and halogens. The multiplication process in a counter may be concluded from the following flow chart (Fig. 4)

There are a number of radiation detectors that operate in different regions of the voltage response curve:

Fig. 4 Multiplication process in the detector

The detector experiences an interaction

↓

The ion pairs are formed

↓

Electrons migrate towards the anode

↓

Electrons reach the region of gas multiplication

↓

Cascade happens

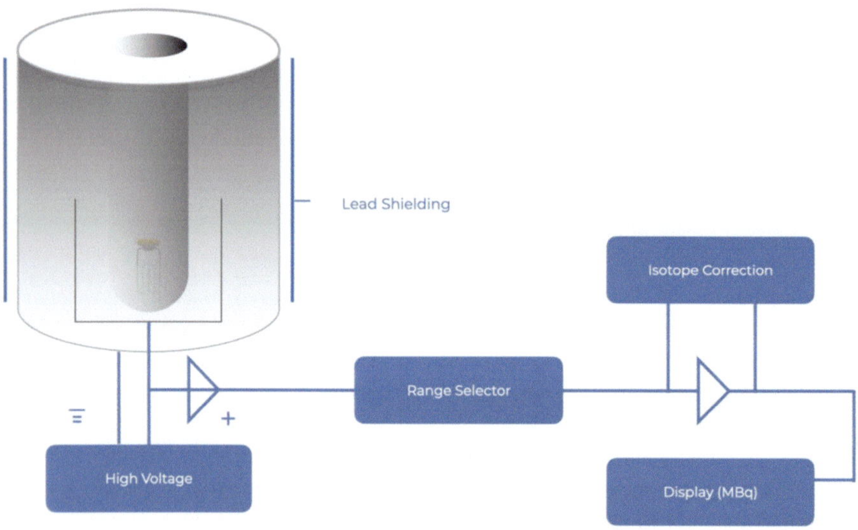

Fig. 5 A dose calibrator

2.2 *Dose Calibrator*

Modern dose calibrators consist of a cylinder into which the radioactive sample is lowered using a sample receptacle. An outer cylinder containing a gas and a positively charged electrode surrounds this cylinder. Electrons are produced by the radiation emitted by the sample and move towards the positive electrode, generating a current that can be measured (Fig. 5). A dose calibrator measures the activity of radionuclide samples (Fig. 5). It is measured in units of Curie (Ci) and/or Becquerel (Bq). It is a true mainstay of nuclear medicine, as the vast majority of departments inspect each and every radiopharmaceutical dose received or dispensed.

2.3 Ion Survey Meter

These are gas-filled detectors mainly used to determine the presence and intensity of radiation. They may also be called "exposure rate meters". An ion survey meter contains a detector and a readout display. The detector has a gas-filled cylinder with two electrodes. When voltage is applied to the device, the gas becomes ionised whenever the counter is brought near radioactive substances. Ion pairs are produced which move towards their respective electrodes, and an electrical signal is produced that is amplified and displayed as a value.

2.4 Ionisation Chamber

The ionisation chamber is the uncomplicated of all gas-filled detectors that are extensively useful in the detection and evaluation of specific types of ionisation radiation, gamma rays, and alpha particles. A very adaptable device is the ionisation chamber. It can be designed in a variety of shapes, diameters, and filling gases.

An electric field is created in the fill gas by applying a voltage between the electrodes. This electric field causes the ionisation of the gas between the electrodes and ion pairs are formed. These ion pairs migrate to the opposite electrodes. This generates an ionisation current, which is measured by an electrometer.

Each ion pair that is created adds or subtracts a small amount of electric charge from or to an electrode. This means that the total charge is proportional to the number of ion pairs made and, by extension, the amount of radiation. This continuous generation of charge creates an ionisation current, that is a measure of the total ionising radiation entering the chamber. However, the chamber cannot distinguish between beta or gamma radiation or generate a radiation energy spectrum.

There are a number of applications associated with ionisation chambers.

- These are extensively used in research labs, the nuclear power industry, radiography, and environmental monitoring.
- Ionisation chambers are routinely used in the nuclear industry because the output is proportional to the radiation dose. As they have a longer operating life than the standard GM tubes, they are widely used in conditions where a constant high dose rate is to be measured.
- In a smoke detector, the surrounding air is allowed to freely enter the ionisation chamber. ^{241}Am, an emitter of alpha particles that generates a steady ion current, is present in minute amounts in the chamber. If smoke penetrates the detector, it breaks this current because ions collide with particles in the smoke and neutralise them. The alarm is set off by this decrease in current.

Ionisation chambers have the advantage that they are exact and precise; approved for beam calibration; instrument readout systems; used in nuclear medicine; and important structures in dose collimators. These chambers only use the discrete

charges produced when the incident radiation interacts with the gas and do not employ the gas multiplication techniques used by other radiation instruments, such as the Geiger-Muller Counter or the proportional counter (which will be discussed later). These chambers have an excellent homogenous response to radiation across a broad range of excited states and are the preferred methods of detecting high concentrations of gamma radiation.
However, connecting cables and a high-voltage supply are required, and many corrections are also needed for high-energy dosimetry.

2.5 *Proportional Counter*

As previously discussed, the basic cylindrical form with a central anode wire and the exterior walls functioning as the cathode is a highly popular design for proportional counters (Fig. 6). Noble gases are the commonly used gases for β measurements. Proportional counters use a slightly higher voltage between the cathode and anode.

As already discussed in the previous section, gas amplification occurs. Because of this gas amplification, the current flow increases, and the number of collected ion pairs is proportional to the number of primary electrons generated by the radiation until a certain voltage is reached, at which point a plateau appears. Avalanche in a proportional counter stops when all liberated electrons get collected at the anode.

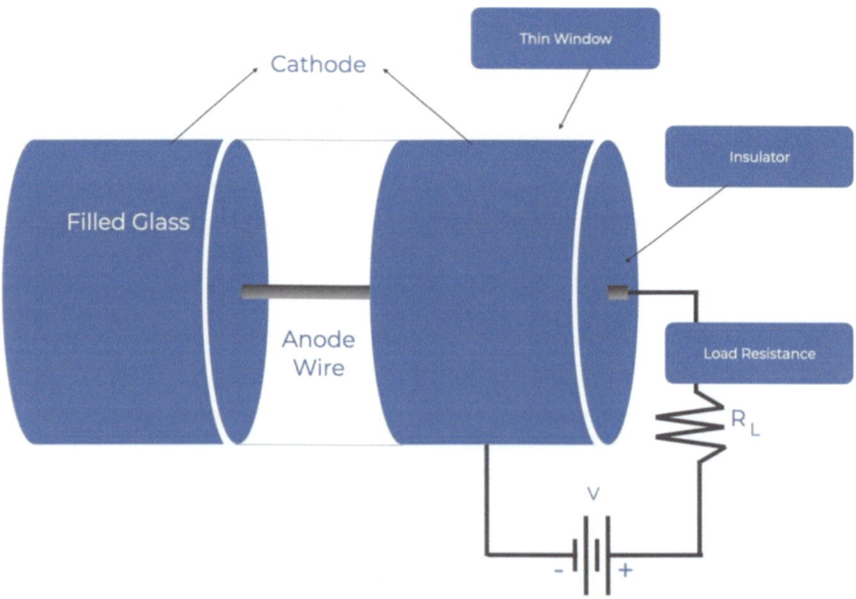

Fig. 6 Basic elements of a proportional counter

This charge amplification in the detector reduces the need for external amplifiers. Gas multiplication needs high electrical fields (~10^6 V/cm).

Proportional counting offers the advantage that particles of varying energies can be distinguished by pulse-height analysis as the size of the current pulse is proportional to the energy of the original charged particle, creating ionisation. Therefore, there is a possibility that one can differentiate between the ionisation produced by heavily and lightly ionising radiations, separating the larger pulses from the smaller pulses produced by alpha particles or gamma rays. Therefore, if a sample is labelled with two dissimilar isotopes, both of these can be detected by these counters. However, counters working in this region need an extremely stable voltage supply, as even small changes in voltage result in huge changes in amplification. Proportional counters are usually only used in a laboratory setting. Scintillation counters can also differentiate between two different isotopes without any of the drawbacks of proportional counters, as will be discussed later in this chapter.

2.5.1 Applications

Spectroscopy Proportional counters are beneficial for charged particle spectroscopy because the energy associated with the charged particle passing along the chamber and the charge created are proportional. By measuring the total charge between the electrodes, one can calculate the particle's kinetic energy, as the number of ion pairs produced by the ionising charged particle is in proportion to its energy.

Photon Detection Proportional counters are also helpful for detecting things like gamma rays, provided they can pass through the entrance aperture.

Radioactive Contamination Detection Proportional counters, that are large-area planar detectors, are widely used to check for radiation dose on personnel, flat surfaces, tools, and articles of apparel. They are typically used to detect alpha and beta particles separately and can also differentiate between different types of radiation by generating an output pulse proportional to how much energy each particle deposits in the chamber. However, their efficacy for beta particles is higher than for alpha particles.

These compartments operate at a negligibly small amount of positive pressure above the ambient atmospheric pressure. They are called "gas-flow proportional counters" if the chamber's gas is sealed or if it is continuously altered. Gas flow varieties have the advantage of being able to tolerate minute holes in the Mylar screen that may occur during use, but they need a continuous gas supply.

For energy measurements of low-energy electrons, the proportional counters are capable of better resolution than the scintillation detectors used for the same application.

2.6 Geiger-Muller Counter

It is one of the earliest and most widely used radiation detectors, invented by Hans Geiger in 1908–1912 and upgraded by Walther Muller in 1928. The counter is named after its inventors. It operates in the voltage-response curve at 400–1000 V (Fig. 3). The generation of an electronic signal in a Geiger counter is more complicated than that of detectors operating in the ionisation region. GM detectors are the most sensitive of the gas-filled detectors. The utility of the GM counter is a result of some of its important characteristics that include high sensitivity, adaptability for use with various types of radiation, high efficiency, a wide variety of shapes and windows, a large size of the output signal, insensitivity to the output signal, and a reasonable price.

A GM counter consists of a copper or brass cylinder (or a glass cylinder which has been silvered) closed with a thin mica window through which the radiation can enter (Fig. 7). The anode wire is thin, such that the charge on it generates an intense electric field energy near to its surface, where the energy is inversely related to the radius of the wire. The anode is kept at a positive potential V of about +400 V relative to the cathode. The cylinder is filled with a noble gas (usually Helium or Argon). The gas is present at low pressure (100 torr) along with a small volume of an easily dissociable "quenching" substance like ethanol or a polyatomic organic gas or alcohol vapour at a pressure of about 10 mmHg.

The radioactive material is placed outside the mica window. Each α or β particle in the tube strikes the gas inside and produces an ion pair. These ion pairs strike other molecules of argon gas and produce more ions and electrons. The ionisation of gaseous molecules facilitates a momentary flow of current. When the voltage is increased further, the proportional region is lost, and the pulse size is no longer proportional to the number of primary electrons or the applied voltage. If the proper operating voltage is given to the tube, electrons in the gas close to the anode and positive ions near the cathode are collected almost quickly. The remaining ion pairs (electrons and ions), as well as the consequences of ion multiplication, follow in

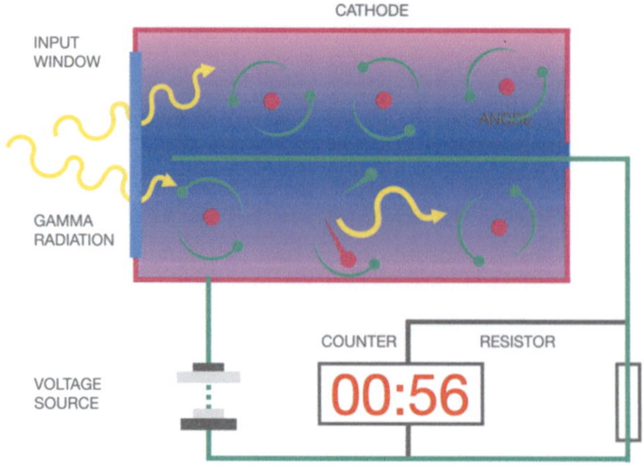

Fig. 7 General plan of a Geiger-Muller counter

quick succession. When the voltage is increased further, ions and electrons are greatly increased. The mechanism is that when the voltage is raised, the electric field moves faster and strikes the gas atoms more violently on their way to the anode for attraction. When the voltage is very high, the electrons collide with the gas atoms with enough kinetic energy to drive electrons out of the atoms. This process is known as "ionisation" by collision, and the atoms become ions that move towards the cathode. The additional electrons unite the original ones in their way towards the anode and at a higher voltage. The ejected electrons are sufficiently accelerated to generate additional ions and electrons. An ion pair generated by ionising radiation is then advanced to high energies close to the wire, resulting in an "avalanche" of additional electron-ion pairs produced by collision with gas molecules. Eventually, a point is reached where the shower of ions appears as a voltage pulse, i.e., electric current flows across a resistor in the high-resistance circuit and the gas is said to breakdown. In the meantime, the positive ions are also drawn towards the cathode, but because they are significantly larger than electrons, they progress comparatively slowly. At a certain point, the anode is totally enveloped by an envelope or "hose" of positively charged ions. Due to the high levels of positive ions and the effective reduction in the electric field, new electrons produced at this point are more likely to combine with a gas molecule than to induce ionisation. The avalanche therefore comes to an end after generating a significant electrical pulse. A pulse takes around 1 μs to be generated and about 50–100 μs to disappear. At a sufficiently high voltage (800–1500 V) between the electrodes, a plateau is reached where the number of pulses of current becomes independent of the voltage and become proportional to the number of ionising particles entering the tube. Most detectors are run at a voltage in the middle of this plateau.

Key Concept
Dead time is the time interval following a pulse during which the counter becomes insensitive to subsequent ionising events.

Typically, ethyl alcohol and ethyl formate are added to the principal fill gas as quenching agents. The positive charge is transferred to the quench gas when positive ions from the primary gas clash with neutral molecules from the quench gas (in fact, an electron is transferred to the positive charge). After neutralisation, the extra energy is mostly used for dissociating the complex quenching gas molecules because all the ions that arrive at the cathode will be those of the quenching gas. Therefore, there is little depletion of the quencher. It takes from about 50–200 μs to quench the discharge. During this time, the GM counter does not respond to any other radiation. Therefore, this is the *dead time* of the detector and is not paralysable. The maximum usable count rate for a typical GM counter is 60,000 counts/min. After reaching the electrodes, the ions are neutralised. Some of these ions, however, might "escape" and produce their self-ionisation avalanche. Hence, if proper care is not done, a Geiger-Muller tube might give a continuous discharge.

Since the termination condition needs a fixed quantity of positive charge, the discharge chain continues till it satisfies the termination condition, irrespective of the initial number of electron-ion pairs produced by the incident radiation. Consequently, all GM counter output pulses have the same amplitude, rendering GM counters unsuitable for radiation spectroscopy.

2.6.1 GM Curve

Figure 8 shows a voltage characteristic curve of a GM tube where count rate due to a constant intensity radioactive source is plotted against voltage across the counter. The GM curve starts with a *starting voltage*. As the voltage is increased, there is a point where the counter starts counting. This is the *threshold voltage* of the counter. After that there is a plateau is reached where there is a small change in counting rate with an increase in voltage. Finally, at a point called the *breakdown voltage*, where the self-suppressing process no longer works, the count rate rapidly rises until there is a continuous discharge in the tube. In order to ensure stable operation, the counter is operated in the plateau region at a voltage corresponding to approximately the midpoint of the plateau.

The measurement units are depicted in the form of contamination units expressed as counts per minute (cpm) or exposure units of mR/h, or both.

The GM tube is extra sensitive to beta radiation. It can detect α radiation of high energy (^{32}P) as well as weak emitters (^{35}C, ^{14}C), but is incapable of detecting ^{3}H because the radiation is unable to penetrate the end window. For a similar reason, the counter is not a very efficient detector of α radiation. The Geiger-Muller counter shows the number of rays per minute entering the tube, but it does not indicate the energy of the radiation.

The basic advantage of the Geiger counter is due to the amplification, i.e., each primary ion pair produces approximately 100 million secondary ion pairs. The

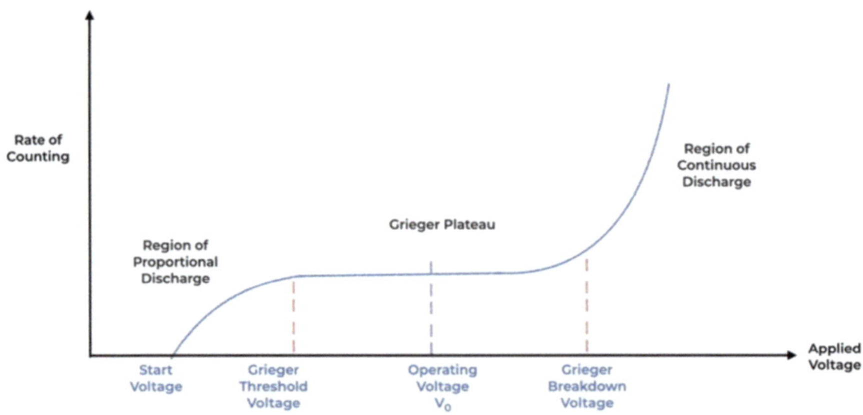

Fig. 8 GM curve

amplification is so large that a very simple external electronic amplifier is sufficient. This simplicity gives stability to Geiger counters. For the detection of α-rays, there is a small window made of thin aluminised Mylar in the GM tube. For the detection of X- or γ-rays, this window is generally closed. Also, because the interaction of γ-rays in the gas volume itself is very less, the detection of these radiations is mainly via the photons or Compton electrons generated in the inner walls of the GM tube. This device is popular both in hospitals using radioactive nuclides and in nuclear generation centres. These instruments are also frequently used in molecular biology studies to monitor the emission of high-energy alpha and beta particles. The output of these compact instruments is also connected into a sound mode, where each "click" represents a particle coming in the tube. One "click" every two seconds can be considered baseline noise.

The GM counter is useful up to 15,000 counts/min and the counter life is limited to about 10^{10} counts. The counting efficiency of a charged particle entering the tube is 100% as even a single ion pair can set off a full Geiger discharge.

2.6.2 Applications

Particle Detection

The first historical application of the GM principle was the evaluation of alpha and beta particles, and this is still the instrument's primary function today. For detecting alpha particles and low-energy beta particles, the "end-window" form of the G-M tube must be used. Because these particles have a small range even in free air and are readily stopped by a solid material, therefore, the tube needs a window that is sufficiently thin to enable more of these particles to pass through to the fill gas. Mica with a density of 1.5–2.0 mg/cm^2 is typically used to make the window.

High energy beta particles are also detected using a thin-walled "windowless" G-M tube (with no end window). Due to their low cost, robustness, and relatively high detection efficiency, particularly with high-energy beta particles, GM counters are useful as portable radioactive contamination measurement and detection instruments. However, for discriminating between alpha and beta particles, scintillation counters or proportional counters are preferred because these instruments are designed with significantly larger detector areas. So checking for surface pollution is easier and faster than with a G-M instrument.

Gamma and X-ray Detection

The windowless tube is used in Geiger detectors, which are frequently used to detect gamma radiation. However, efficiency is typically low because gamma rays interact with matter less strongly than both alpha and beta particles.

For high-energy gamma rays, interaction of the photons with the tube wall, typically 1–2 mm of steel on a "thick-walled" tube, produces electrons in the wall that can penetrate and ionise the fill gas. This is necessary because the low-pressure gas in the conduit has very little interaction with high-energy gamma photons. However, for photons with minimal energy, there is more gas interaction, and the direct ionisation consequence of the gas increases. As energy decreases, the wall effect is replaced by a combination of the wall effect and direct ionisation, until direct gas ionisation predominates. Due to the variation in response to various photon energies, windowless tubes use "energy compensation," which tries to compensate for these variations over a broad energy range.

Neutron Detection

Neutrons are measured using a variation of the Geiger tube with boron trifluoride or ^{3}He as the gas and a plastic stabiliser to delay the neutrons. This produces an alpha particle in the detector, allowing for the counting of neutrons.

Gamma Measurement: Personnel Protection and Process Control

The principal use of GM detectors in nuclear medicine is in radiation protection work for surveying the contamination of surroundings and personnel from X- and γ ray emitting radionuclides. A Geiger tube is still being used as a sensing device; however, the processing electronics will be more sophisticated and reliable than those used in handheld survey meters.

They are also useful for rapid screening of radioactive gels before performing autoradiography or for examining the chromatographic fractions for labelled components.

In spite of all these applications, GM counters suffer from several disadvantages, which are listed below:

1. The necessity of having high voltage,
2. Inability to distinguish pulses because of physical processes, i.e., they are not able to differentiate radiation types,
3. Cannot determine the energy of different radiations accurately, and
4. Efficiency is very low.

Table 2 showing the comparison of various aspects of different gas-filled detectors is depicted below.

Due to their numerous benefits, gas-filled radiation detectors are still utilised in a variety of applications. Their application in photo-spectroscopy is restricted to energies below 200 keV. Even in battery-powered portable instruments, sophisticated low-power microelectronics for signal processing in these detectors extract useful information.

Table 2 Comparison of various types of gas detectors

S. No.	Parameters	Ionisation chamber	Proportional counter	GM counter
1	Mode of operation 1. Pulse 2. Current	Possible in typical current mode	Possible in typical current mode	Typically pulse frequency Non-linear current response to dose rate
2	Spectrometry	Yes, for alpha particles	Yes, Gamma rays up to 200keV,	Not possible because pulse height is independent of energy
3	Gas amplification	1	10^3–10^5	$>10^6$
4	Types of radiation detected	$\alpha, \beta, \gamma, X, n$	$\alpha, \beta, \gamma, X, n$	α, β, γ, X α with thin window
5	Types of gas/ pressure	Air or any other gas/≥ atmosphere	Counting gas quenching gas/ depends on application	Counting gas Quenching gas/<atmosphere
6	Order of pulse height	<10 mV	≤100 mV	≈ Tens of Volts
7	Electronics	Care for small pulse amplification and low current measurement	Stable EHT, moderate complexity electronics	Simple electronics, less critical operating conditions
8	Applications	Source calibrators, area monitors, β/γ dosimeters, for measurement of radioactive gases, etc.	Surface contamination windowless counting for nonpenetrating radiation, position sensing	Area monitors, personal monitors, counting applications, contamination monitors, etc.

Check Points

- What are gas-filled detectors?
- Describe the general assembly of gas-filled detectors.
- Explain the voltage response curve of gas-filled detectors.
- What is a proportional counter? Explain its working.
- Write the applications of the ionisation chamber.
- Explain the working of a GM counter.
- Write down the advantages and disadvantages of the GM counter.
- Describe the voltage response curve of a GM counter.

Therefore, it can be concluded that detectors based on gas ionisation are ionisation chambers, proportional counters, and Geiger-Muller counters. All these detectors have a gas chamber where the radiation enters, causes ionisation of the gas, and leads to the production of an electric current that is measured. These instruments are involved both in detecting the presence of ionising particles as well as in measuring ionising radiation. All the three detectors of this category have the same basic design. However, the discrimination between different kinds of radiation is possible with an ionisation chamber and a proportional counter.

3 Methods Based on Excitation

Gas ionisation-based counters used to be the chief instruments employed to quantify radioisotopes in biological samples. However, due to some drawbacks associated with the gas-filled detectors, as discussed previously, these have now been replaced by scintillation counters.

Radioisotopes interact with matter in two ways:

- Ionisation, which is from the basis of GM counting (already discussed), and
- Excitation

The latter effect leads to an excited molecule (called flour) that emits photons of light. *This phenomenon of fluorescence because of excitation by radioactivity is known as scintillation.* Scintillations are actually tiny flashes of light that are produced by certain materials when they absorb radiation. These substances are known as *fluorescent materials*, *fluors*, *scintillators*, or *phosphors*. A *scintillator* is any substance that can emit a photon in the ultraviolet or visible light spectrum when an electron in the scintillator returns back to its ground state. The photons pass through the light tube and fall on a photosensitive material known as the *photoelectric cathode* surface of the photomultiplier tube. The electrons are released from the photoelectric cathode surface. These electrons are focused towards an electrode (called *dynode*) at a faster rate. The surface is coated with a material that generates secondary electrons four or five times greater than the incident ones. The secondary electrons then collide with the second dynode after further acceleration, thereby multiplying the number of electrons further and continuing along the tube (Fig. 9). There are a series of such electrodes (dynodes) for the acceleration and multiplication of electrons. A single ionising particle can produce millions of electrons in a PMT. These enormous number of electrons strike the collector plate to generate a current, which is amplified further and electronically recorded. These scintillation photons can be detected and quantified by a photomultiplier tube when the electron comes back to the ground state and releases electromagnetic radiation. In the case of gamma radiation, scintillation is caused by the excitation of the bound electrons by free electrons within the scintillator. These free electrons are produced by any of the three interactions, i.e., the photoelectric effect, the Compton effect, and pair production. The probability of the occurrence of these interactions depends on the type of scintillators and the energy associated with the gamma radiation. This is the

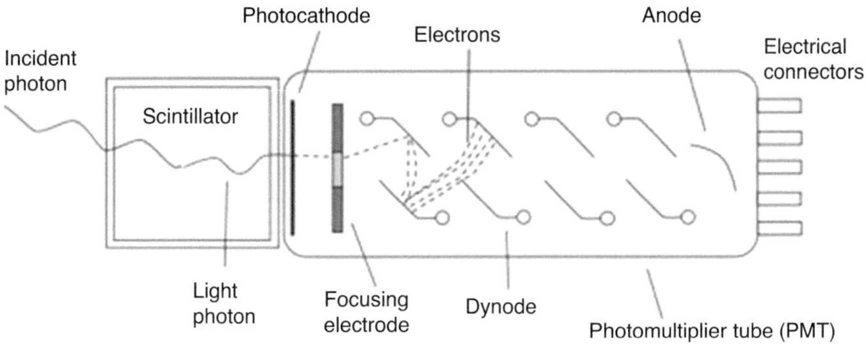

Fig. 9 Scintillation counter. Adapted from Scintillator Materials Group. Stanford University. Robert S Feigelson

basis of *scintillation counting*. The electric discharge that is generated from the photomultiplier's conversion of light energy to electrical current is directly proportional to the energy of the initial radioactive event. Also, the number of scintillations is proportional to the rate of decay of the sample, i.e., the quantity of radioactivity, as well as the intensity of light emitted. Consequently, the signal from the photomultiplier is proportional to the energy of the radiation. The counter could therefore be used for both qualitative and quantitative applications. This is a significant advantage of scintillation counting because it enables the detection and measurement of two or more isotopes in the same sample, provided that their emission energy spectra are adequately distinct. This is the main advantage of scintillation counters over Geiger counters.

The instrument that is used for detecting scintillations is called scintillator counter. It has various advantages over gas-filled detectors. Because they are solid rather than gaseous, they interact with gamma rays with much greater efficacy than detectors filled with gas. This provides a way to measure the energy of a radiation interaction. Low-energy alpha and beta particles may not be detectable with GM counters, so scintillation counters are used in the laboratory to detect these emitters. Tritium (^{3}H) can only be found using a scintillation counter.

Scintillation counters are frequently used for bioassays, to detect spillage using samples from suspected contaminated areas, and to monitor finger exposure in laboratory personnel. Scintillation detectors have a wide range of applications in numerous processes involving the detection of gamma rays. In nuclear medicine, they are found in thyroid instruments, well counters, gamma cameras, and positron emission tomography (PET) systems. Specifically, solid and liquid scintillation measurements are standard laboratory methods in the life sciences for measuring radiation from gamma and beta emitting nuclides, respectively. This methodology is routinely used in the overwhelming majority of diagnostic and/or research laboratories, from biochemistry and biology laboratories to clinical departments.

3.1 Scintillation Counter

A scintillation counter is a device that counts the radioactivity and also determines the energy of the radiation. Scintillation counting provides two types of information:

(a) The light intensity produced and the signal obtained from the photomultiplier are directly proportional to the original radioactive energy, and this gives a *qualitative assessment.*
(b) The number of scintillations created is proportional to the rate of decay, and therefore the amount of radioactivity is *quantitatively measured.*

Scintillation counters consist of a scintillator and the associated electronics, that include a photodetector that can be either a photomultiplier tube or a photodiode (Fig. 9)

3.1.1 Scintillator

Scintillators form the most important component of scintillation counters. A scintillator is a material that emits fluorescence when exposed to ionising radiation such as gamma rays. It is a type of phosphor. When radiation collides with the scintillator, it absorbs its energy, and the intermolecular electrons move from the ground state to the excited state. During their transition to the ground state, near-visible or visible light is emitted. Many different types of materials have the capability to scintillate. Scintillation materials come in a variety of physical forms, including crystals, glasses, granules, ceramics, plastics, liquids, and gases.

Different types of scintillators used are discussed below.

Organic Scintillators

These are aromatic hydrocarbon compounds containing interconnected benzene ring structures in a number of ways. Some organic scintillators are pure crystals made of anthracene, stilbene, and naphthalene. Their luminescence decays within a few nanoseconds. These scintillators have a quick decay time and very durable. However, their response is anisotropic (which degrades energy resolution if the source is not collimated), they cannot be readily machined, and they cannot be grown to large sizes; as a result, they are rarely used.

Organic Liquids

These are solutions with one or more organic scintillators mixed in an organic solvent. Examples include p-terphenyl, PBD, butyl PBD, and POPOP. The most widely used solvents are toluene, xylene, benzene, etc. These are mainly used in liquid scintillation counters. To match the spectral sensitivity range of a specific PMT, liquid scintillators are additionally supplied with additives such as wavelength shifters.

Plastic Scintillators

This is another type of scintillating material in which the primary fluorescent emitter, known as a *fluor*, is suspended in the base, which is a solid polymer matrix, such as polyethylene naphthalate. The most common bases used in plastic scintillators are aromatic plastics, among them are polyvinyl toluene (PVT) and polystyrene (PS). The benefits of plastic scintillators are a relatively more light output, easy handling and management, a faster signal with a decay time of 2–4 ns, the ability to be moulded into almost any desired shape, and high durability. They are also easy to handle.

Inorganic Scintillators

These are crystals formed in high-temperature furnaces, such as alkali metal halides, that frequently have a trace amount of activator impurity. Scintillation crystals must be optically transparent, with no fissures or boundaries that could cause photons to be reflected. Scintillation crystals are extremely fragile and can get fractured under mechanical stress or rapid temperature changes (>5 °C or 9 °F/h). As a result, extreme caution must be exercised when operating near an exposed crystal. These scintillators are used in nuclear medicine applications, frequently with trace amounts of impurities to improve their scintillation performance. An example includes thallium-activated sodium iodide, or NaI (Tl), the most widely used inorganic scintillator in nuclear medicine, developed by Robert Hofstadter in 1948 for radiation detection. The fact that the same crystal is used for different samples, which improves reproducibility, is an advantage. Compared to other scintillators, NaI (Tl) crystals are efficient and produce intense pulses of light. However, due to their low resolution, they are unsuitable for identifying complex mixtures of gamma ray producing substances. NaI is hygroscopic, due to which the crystal turns yellow and absorbs the scintillation photons rather than transmitting them. Scintillation crystals must be hermetically sealed to prevent contact with oxygen or water. CsI(Tl), CsI (Na), and CsI are additional inorganic alkali halide crystals. BaF_2, CaF_2 (Eu), ZnS (Ag), etc. are non-alkali crystals.

Inorganic crystals can be easily cut into minute pieces and arranged in an ordered configuration to provide position sensitivity. In medical physics and security applications, such arrays are frequently used to detect x-rays or radiation.

Gaseous Scintillators

These scintillators consist of nitrogen and the nobel gases such as He, Ar, Kr, and Xe. The scintillation process occurs due to the de-excitation of atoms that were excited by the passage of an incident particle. The de-excitation is very rapid (~1 ns), so the detector response is very quick. In nuclear physics, these detectors have been used to detect fission fragments.

Glass Scintillation

The most usual glass scintillators are cesium-activated lithium or boron silicates. These are particularly suitable for the detection of thermal (slow) neutrons. Li is more widely used than B since it has a greater light output. Glass scintillators are, however, sensitive to electrons and γ-rays as well. They are well suited to harsh environmental conditions. Their response time is approximately 10 ns.

3.1.2 Associated Electronics

Photomultiplier Tube

The amount of light produced in a crystal is quite small relative to that which the human eye can easily detect. Even if one could observe each scintillation occurring in a scintillator with the naked eye, it is not practical to count light flashes. Therefore, the scintillation photons are to be converted into an electrical signal. This is accomplished by a PMT. A scintillation crystal is attached to a photomultiplier tube (PMT), that is a light-sensitive device that converts light released into measurable electronic pulses. The PMT contains a photocathode facing the window through which radiation enters, a focusing grid, an array of dynodes arranged in a special geometric pattern, and an anode placed in an evacuated glass tube. The photocathode is made of a photosensitive material that emits electrons when exposed to light photons produced by radiation in the scintillation crystal. Metal alloys with additional electrons are typically used as dynodes in PMTs, usually bialkali antioxide compounds such as K_2CsSb and Na_2KSb. This photoelectron is then accelerated towards a dynode by the *focusing grid* by the application a voltage (between 50 and 100 V) to that dynode. The grid focuses these electrons onto the first dynode. As a result of this acceleration, the electron acquires sufficient kinetic energy to produce a number of secondary electrons. These secondary electrons are then accelerated towards a second dynode (at a voltage of two times higher than the voltage applied on the first dynode), where further multiplication of electrons occurs. Eventually, there will be between 10^5 and 10^8 electrons for each of the photoelectrons produced. Therefore, an array of *dynodes* multiplies the smaller number of electrons produced by the photocathode into a detectable electronic signal. Finally, the anode accepts and collects all the electrons, and an electric current is produced (Fig. 9). The crystal and the PM tube are generally combined into a single, light-tight unit. This prevents ambient light from reaching the PM tube.

Preamplifier

The electric pulses appearing at the anode of the PM tube are still small (μA) and require amplification before they can be further analysed or processed. The output of a PM tube, however, cannot be directly fed to an amplifier because of the wide difference in the output impedance (an important electronic parameter) of a PM tube and the input impedance of an amplifier, that results in single distortion and attenuation. A preamplifier is a device that solves this impedance mismatch. It is located next to the PM tube (to minimise signal distortion). The PMT output signal is amplified by a preamplifier so that it can be processed by the remaining detector electronics.

Linear Amplifier

A linear amplifier amplifies the incoming pulses in direct proportion. Multiplication factors in amplifiers are usually between 1 and 100 and are modified using the coarse and fine gain settings (the energy multiplication knob) on the electronics module. In order to produce a series of separate pulses that are simpler to analyse, the amplifier also shapes and shortens the pulse.

Pulse-Height Analyser (PHA)

A PHA is an electronic device in the counter that accepts the pulses whose voltage (height) lies within a preselected range while rejecting the pulses whose voltage lies outside this range. It dictates the size of the pulse, that is proportional to the energy of the gamma ray. There are two PHAs: the single channel analyser (SCA) and the multichannel analyser (MCA). An SCA is provided with two controls known as a lower level discriminator (LLD) and an upper level discriminator (ULD) that only allow voltage pulses lower than their setting. The LLD and ULD values are specified by the operator and are dependent on the measured photons' energy. Each pulse is compared to the LLD and the ULD, and pulses having an amplitude greater than the LLD but less than the ULD produce PHA output pulses. All PHA output pulses of similar size and all go to the scalar and rate meters. The SCA mainly pulls the gamma ray energy data from the voltage indicator and, based on that energy, admits or rejects each pulse. The primary disadvantage of a SCA is that it produces an output pulse only when the input pulse falls within this range, i.e., between the LLD and the ULD; consequently, a significant quantity of potentially useful information is lost.

The multi-channel analyser (MCA) is a circuit that can set up a large number of separate windows to examine the entire spectrum at once. In a multi-channel analyser, there are many PHAs (1000 or more), each of which represents a little portion of the energy spectrum that can simultaneously separate the pulses into multiple ranges. Each gamma ray is documented as a "1" in one PHA and a "0" in all other PHAs. Therefore, the whole spectrum of pulse-height distribution can be measured once. A computer is attached to it that can simultaneously get data from each window and shows it in the form of an energy spectrum. The computer software permits the manipulation of the resultant data in a number of ways.

Scalar and Timer

These devices comprise an electronic counter or digital display that is used for counting the pulses coming from an amplifier or SCA either for a specific time interval or until a predetermined number of counts have been collected. In either case, both the number of counts and the time during which they are collected are obtained.

Rate Meter

The rate meter is a display device that indicates the number of pulses produced per unit time. This output voltage (V_{out}) of the scintillation detector is directly proportional to the energy transferred to the crystal by the radiation. Rate meters are now rarely used in nuclear medicine instruments.

The overall scheme of operation of a PM tube is depicted in the flow chart.

Light photons

↓

Ejection of an electron (photo-electric event) from a *photocathode* (at 0V) at the front of the PM tube

↓

Acceleration of this electron to the first positive potential *dynode*

↓

Kinetic energy of the single impacting electron causes further recoil electrons

↓

Acceleration of these electrons to the next dynode (held at a higher positive potential than the first dynode)

↓

Creation of several electrons for every impactinge lectron

↓

Finally ~ millions of electrons collected at the final positive potential anode

↓

Output current

↓

Output Voltage pulse

Check Points

- What is a scintillator?
- State the different types of scintillators.
- How does a scintillator lead to the detection of radiation?
- Explain the role of the electronics in the scintillation counter.

Thus, scintillation counters measure radiation by detecting flashes of light or scintillations that are produced in certain materials by the radiation. These scintillations are converted into electrical pulses by suitable electronics attached to the counter. Scintillators are made of plastic, organic, and inorganic materials and they can be solid, liquid, or gas. Scintillation counters are able to discriminate between different types of radiation and even between different energies of the same radiation.

3.2 *Types of Scintillation Counting*

There are two methods of scintillation counting, depending upon the nature of the sample under consideration:

- Solid scintillation counting,
- Liquid scintillation counting

Key Concept
While working on the Manhattan Project at the University of California, Berkeley, Sir Samuel Curran invented the first electronic scintillation counter in 1944.

3.2.1 Solid Scintillation Counting

Here, the sample is positioned next to a crystal of a fluorescent material (Fig. 9). The transparent inorganic crystal, called a scintillator, emits fluorescent light when irradiated by the sample. Thallium-doped sodium iodide (NaI) (Tl) crystal is the most commonly used crystal. This crystal consists of different sizes for various types of equipment. The radioactive sample is placed in a scintillation vial made of glass or plastic, which is then deposited directly on a solid scintillating material, desiccated, and counted on a scintillation counter. The crystals are positioned close to a photomultiplier, which is connected to a high voltage power supply and a scalar.

Solid scintillation counting is mainly useful for detecting γ emitting isotopes. This is because γ-rays are highly penetrating electromagnetic radiation, and they rarely collide with surrounding atoms to cause ionisation or excitation. They interact with the detector via the photoelectric effect, the Compton effect, and pair production methods, and resulting in the generation of light or scintillations throughout the crystal. Atoms in a crystal are densely compressed, increasing the likelihood of collisions. Solid scintillation counting is unsuitable for detecting weak beta-emitting isotopes such as ^{3}H, ^{14}C, and ^{35}S. Even, α radiation is tremendously reduced by even 0.05 mm of aluminium or copper, and so even with the narrowest shielding, it cannot reach a scintillator crystal. Solid scintillation counting is also frequently used in

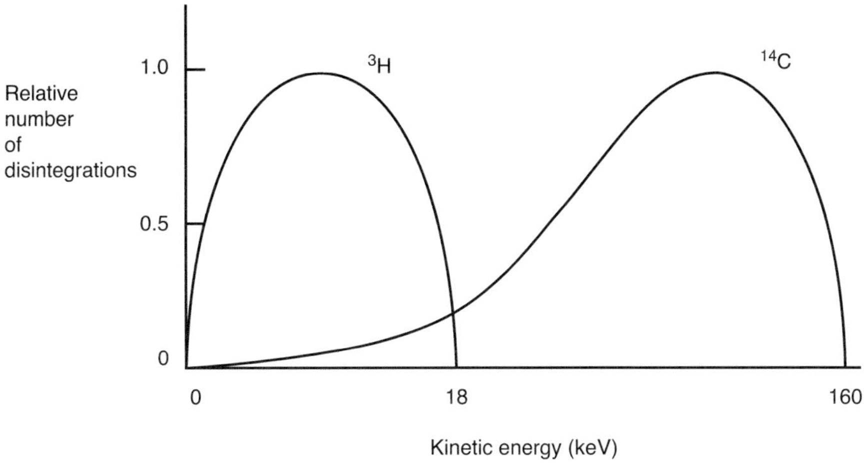

Energy spectra for two commonly used beta emitters. Note that the horizontal scale is not linear.

Fig. 10 Energy spectra double isotope analysis. Adapted from Detection and Measurement of Radioactivity. Experimental Biosciences. By David R Caprette. 1995. http://www.ruf.rice.edu/~bioslabs/methods/radioisotopes/rad1.html)

biological research, for example, in radioimmunoassays where gamma emitters are used.

A scintillation counter makes it possible to measure two different isotopes in one sample. This is called double isotope analysis. However, it is important that the energy spectra of the two isotopes being evaluated differ significantly from each other. There are many such sets of isotopes whose energy spectra are very different. Some of the examples are ^{32}P and ^{14}C, ^{3}H and ^{35}S, ^{3}H and ^{32}P, ^{35}S and ^{32}P, and ^{125}I and ^{131}I (these are two isotopes of the same element). The figure below illustrates the basic principle of double isotope analysis (Fig. 10).

It is evident from the figure that two isotopes (^{3}H and ^{14}C) always have different, non-overlapping energy spectra. For dual isotope analysis, a two-channel apparatus should be used: ULD and LLD. The two channels will measure the isotopes ^{3}H and ^{14}C simultaneously.

3.2.2 Liquid Scintillation Counter

In liquid scintillation counting (also called internal counting), the radioactive sample is mixed in a scintillation cocktail consisting of the solvent and a suitable scintillator (primary and secondary flours) (Fig. 11).

Liquid scintillation counting is extensively fruitful for quantitating soft β emitters.

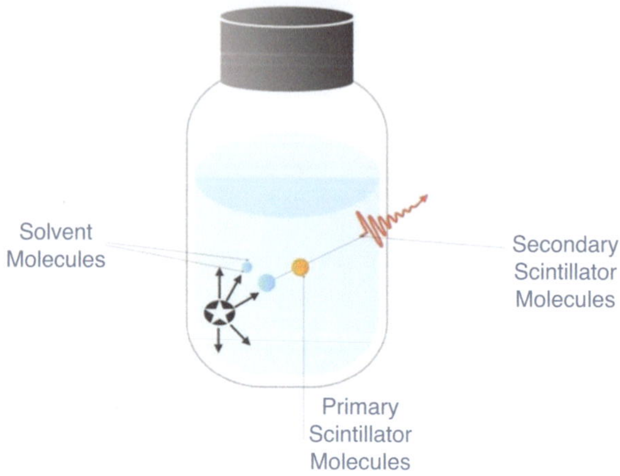

Fig. 11 Liquid scintillation counter

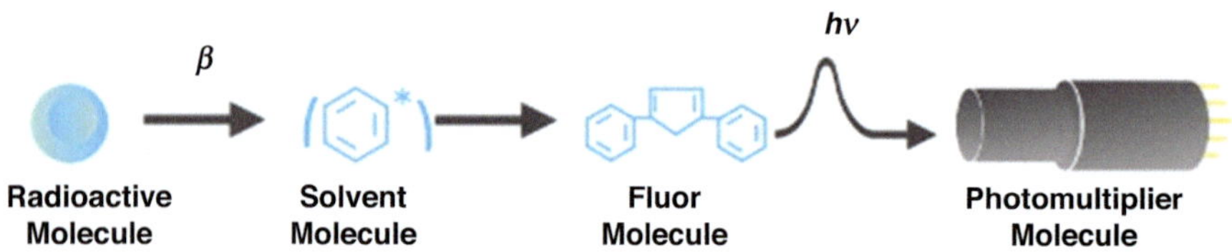

Fig. 12 Energy transfer in liquid scintillation counting

This is the frequent technique for the quantification of low energy beta emitters. Such emitters (for example, ^{3}H, ^{14}C, ^{35}S, ^{45}Ca, etc.) are challenging to detect with portable survey instruments because the beta particles cannot pass through the fine window of the detector.

Beta particles are emitted during radioactive decay in liquid scintillation counters. To ensure efficient energy transfer between the beta particle and the fluors in the solution, the sample material is dissolved in the solution. In the scintillation fluid, the beta particle traverses only brief distances before its entire kinetic energy is dissipated. Typically, it only takes a few nanoseconds for a beta particle to lose all of its kinetic energy. The medium absorbs the energy in three different ways: heat, ionisation, and excitation. Some of the energy is absorbed by solvent molecules, which causes them to become agitated (not ionised). The excited solvent emits ultraviolet light, and the solvent molecule returns to its ground state. The excited solvent molecules transfer their energy to the primary fluor (Fig. 12) by disrupting the electron cloud of the fluor, causing it to become excited. A photon of ultraviolet light is produced as the orbital electrons of the primary fluor molecule revert to their ground state. The UV light is absorbed by the second fluor molecules, which, upon returning to the ground state, produce blue light flashes. Approximately 10 photons per keV of energy are produced during nuclear decay events. The energy is dissipated in around 5 ns. The beta particle's initial energy determines the intensity of the light.

The photomultiplier tube's photocathode is illuminated by a flash of blue light (PMT). Therefore, electrons are released, producing an electrical signal that is proportional to the number of blue-light photons. A very small quantity of light is produced by the fluors. Hence, a sensitive method of detection is needed to measure this event. Many of the commercially available instruments utilise at least two photomultipliers. The signal from each PMT is fed into a circuit that only generates an output when two signals occur simultaneously or within the circuit's resolving time (approximately 20 ns). By aggregating the size of the pulses obtained from each PMT, an output is obtained in proportion to the total scintillation intensity. The magnitude of the electrical discharge is converted to an automated value that represents the beta particle's energy. This is transmitted to the analyser, where it is compared with the digital values for each LSC channel. A cathode ray tube shows the number of pulses from each channel. In this way, the sample is analysed, and the spectrum can be obtained to get information about the radiation's energy or the quantity of radioactive material dissolved (Fig. 12).

Ionising radiation directly or indirectly excites flour molecules and results in the release of photons that fall on the photocathode of a photomultiplier tube to produce photoelectrons from the photocathode surface (Fig. 12). These photoelectrons pass through a dynode series, resulting in an extremely amplified electron pulse at the anode.

3.2.3 Scintillation Cocktail

Thus, it was found that in liquid scintillation counting, the alpha or beta emitter to be analysed is suspended in an organic solvent along with one or more fluors which emit a scintillation of light that can be detected. Thus, there are two major components of all scintillation cocktails: a solvent, that dissolves the sample, and fluor, that emits light. The solvent makes up 66–99% of the cocktail by volume.

Solvent

In a liquid scintillation counter, the radiation from the suspended sample molecules interacts with a solvent molecule. The energy from the sample is transferred to the more plentiful solvent molecules. This transfer of energy is most efficient when using aromatic solvents. A number of solvents with varying relative efficiencies of energy transfer are given in Table 3.

When a sample containing a radioisotope dissolved in a cocktail goes through an emission event, it is very likely that the particle from the sample will only interact with solvent molecules before its energy is depleted. For this reason, the solvent must function as an efficient energy collector and conduct that energy to the phosphor or fluorescent molecules, as opposed to dissipating it through some other mechanism. The solvent must not stop the phosphor's scintillation, and it must

Table 3 Relative counting efficiency of some solvents used for liquid scintillation counting

S. No.	Solvent	Relative counting efficiency (%)
1	Ethanol	0
2	Acetone	12
3	Ethylene glycol	60
4	1,4-dioxane	70
5	1,3-dimethoxy benzene	80
6	Xylene	97
7	Methoxy benzene (anisole)	100
8	Toluene	100

dissolve the phosphor to create a stable, countable solution. As can be seen in the table, aromatic organics have proven to be the most effective LSC solvents. Toluene is a prototypical LSC solvent.

The collision with the solvent molecule and the accompanying energy loss are repeated events which continue until the radiation has shed the majority of its energy. Only around 5% of the total incident energy is finally absorbed as light. The remaining energy is liberated as heat. The number of collisions that the solvent molecule experiences, and therefore the quantity of light emitted, is proportional to the initial energy of the radiation.

Fluor

The efficiency with which energy is transferred between an excited solvent molecule and a fluor molecule depends on their concentrations. Most fluors are oxazole and oxadiazole compounds. Of these, PPO (i.e., 2,5-biphenyl oxazole) is the most frequently used primary scintillator, and POPOP (1,4-(di-2-(5-phenyl oxazolyl))-benzoyl) is the most often used secondary scintillator. Naphthalene can also be used, but its efficiency is low.

The solvent molecule, that gets excited when it collides with the radiation, emits light when it comes back to its ground state. This process is called *phosphorescence*. This phosphorescence is due to the special chemical nature of the solvent, i.e., it has loosely bound π- electrons and can be excited easily by radiation. The solvent releases light of a very short wavelength, which falls in the range of 260–340 nm. This range is so short that it cannot be detected by most existing instruments. To avoid this problem, a second molecule is used. This compound, known as a primary flour (also called a scintillator), absorbs light in the wavelength range of 260–340 nm (this is the emission spectrum of the solvent) and emits light of a longer wavelength (340–430 nm). However, older instruments are insensitive to even this wavelength range and require an even longer wavelength of light. Such a shift to a higher wavelength can be easily obtained by using a secondary flour. The secondary flour absorbs the light emitted by the primary flour and re-emits it at a still longer wavelength. The purpose of introducing this fluor is that the emission spectrum of the

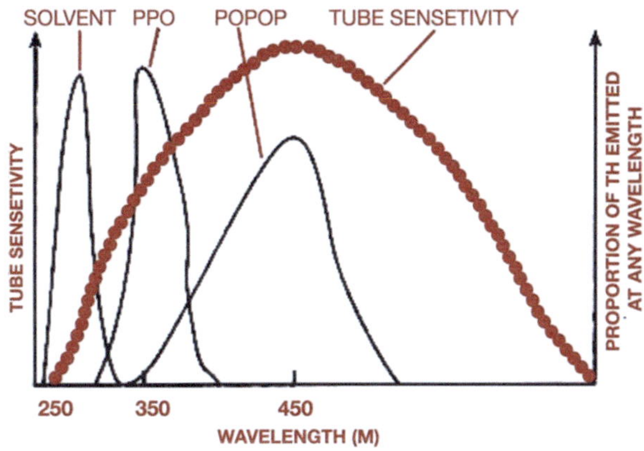

Fig. 13 Role of primary and secondary flours in detection of radiation

sample and the spectral response of the photocathode, which detects the light, complement each other (Fig. 13).

Although PPO and POPOP are among the most frequently used primary and secondary flours, respectively; other compounds such as 2-(4-t-butyl phenyl)-5-(4-biphenyl)yl–1,3,4-oxadiazole (BUTYL-PBD) are better primary flours but are quite expensive and are also affected by extremes of pH. Other examples of secondary fluors include dimethyl POPOP [1,4-Bis-2(4-methyl-5-phenyloxazolyl)-benzene] and bis-MSB [p-Bis (0-methyl-styryl)-benzene].

All the components of the scintillation cocktail as discussed above are mixed and put in a container called *scintillation vials* or *counting vials*. A proper choice of counting vial can influence both the accuracy and reproducibility of the measurement. There are mainly two types of materials used in counting vials: glass and polyethylene. Glass has the property of being unaffected by the solvents which is not the case for polyethylene. In such polyethylene vials, there will be leakage of the solvent and solute through the walls. This implies that repeated measurements of the same sample might give different results. However, the transmission of scintillation light is better in polyethylene vials than in glass vials, mainly due to the reflection of light and less transmission of short waved light in the glass vial. On the other hand, the background is generally higher in a glass vial due to the natural radioactivity present in the glass. Thus, both of these materials have their own advantages and disadvantages. Adsorption of the radionuclide to the walls of the vial can be a problem in glass vials.

Counting Efficiency

Counting efficacy is the proportion of emission events that generate a detectable photon pulse. In general, counting efficiency is the ratio of counts per minute (cpm) to disintegrations per minute (dpm) and is indicated as a percentage. Counting efficiency is dependent on the type of radioisotope, sample composition, and type of scintillation counter. Poor counting efficiency is due to an extremely low energy-to-light conversion rate.

Quenching

Quenching is the decrease in counts caused by the properties of the sample or cocktail and its components. Quenchers may be chemical quenchers or colour quenchers. Chemical quenchers absorb energy before it is transformed into light. Thus, chemical quenchers decrease the number of photons produced by the β particle.

Colour quenchers absorb light in the scintillator's wavelength spectrum. In this instance, the number of photons emitted does not change, but the number that reach the photomultiplier tube does. In both types of quenching, the energy of all light pulses is reduced, and the total cpm is reduced to undetectable levels by the number of pulses quenched. This causes an underestimation of the total counts and, consequently, the isotopes present. It also causes a noticeable shift in the energy spectrum of the sample.

3.3 *Applications*

Liquid scintillation counting offers a number of applications:

3.3.1 Quantitative Study

In a liquid scintillation counter, the intensity of radiation produced is given by disintegrations per minute (dpm). The count rate is determined by how effectively nuclear decay events generate light bursts that are detected and measured by a liquid scintillation counter. Because the scintillation solution is always used, it can absorb nuclear decay energy, preventing it from being absorbed by chemical fluors. This results in the phenomenon known as *quench*, which is defined as interference with the transformation of decay energy into photons released from the sample container. To compensate for quenching, the counting efficiency must be known:

$$\%\text{Efficiency} = (\text{cpm} / \text{dpm}) \times 100 \tag{2}$$

The greater the effect of the quencher on the counting efficacy of the sample, the lower the radionuclide's radiation energy. Techniques for undertaking these corrections have been devised, and significant research has been done to improve the counting efficiency by employing different detection systems: internal standardisation, sample spectrum quench correction, and external standard quench correction.

The number of scintillations is proportional to the sample's rate of decay, or radioactivity. Samples from enzyme inhibition, cytotoxicity, immunoassay, receptor binding, and numerous metabolic studies can be counted using solid scintillators.

3.3.2 Qualitative Study

The prominent benefit of liquid scintillation counting has been the detection of low-energy beta emitters, although it can be the method of choice for other isotopes in certain cases. The quantity of light emitted and therefore the signal obtained from the photomultiplier depend on the energy of the radiation.

Solid and liquid scintillation counting is used for the identification of radioisotopes in diverse areas like biomedicine, ecology, and industry.

A liquid scintillation counter is routinely used for both *in vivo* and *in vitro* biomedical research. The utility of radioisotopes in research derives from their chemical identity as compared to their non-radioactive counterparts. This enables their incorporation into "tracers," which are radiolabelled components that can be followed and detected through a succession of reaction steps. Some of these studies include assays for carbohydrate and lipid metabolism, enzyme activity determination, amino acid and nucleoside transport studies, hormone determination in blood tests, and identification of various radionuclide contaminants in soil samples.

- Other applications include the detection of radioactive contaminants for environmental monitoring and ecology research, such as the investigation of microbial activity.
- Using a method called Cherenkov counting, some beta emitting isotopes can also be analysed on an LSC without the use of a cocktail or with only a small amount of water.

3.3.3 Environmental Liquid Scintillation Counting

Environmental liquid scintillation counting (LSC) is the measurement of both natural and anthropogenic radionuclides in the natural environment. Numerous radionuclides are routinely measured at natural environmental levels in waters, sediments, soils, the air, etc. These include radium (Ra), uranium (U), thorium (Th), ^{210}Pb, ^{222}Rn, and ^{231}Pa isotopes. Examples include ^{14}C dating, groundwater movement and dating using ^{3}H, marine sediment mingling, productivity, and particle flux investigations using ^{234}Th and ^{210}Pb/^{210}Bi/^{210}Po. Incubation with $NaH_{14}CO_3$ can also be used to determine the photosynthetic activity of microorganisms in environmental samples.

- Alpha particle counting by liquid scintillation has a high counting efficiency (nearly 100%). This is crucial to the management of radioactive waste, process control, the nuclear fuel cycle, and environmental protection. Nonetheless, the energy resolution is very weak, so they are not much help in identifying these radionuclides.

- LSC is the most important benefit of scintillation counting for the detection of beta emitters, with performance approaching 100% for high-energy beta emitters (^{89}Sr, ^{90}Y, and ^{32}P) and 50–60% for low-energy emitters (50–60% for ^{3}H, 95% for ^{14}C). Nonetheless, it is frequently the method of choice for these weak beta emitters.

3.4 Advantages

Liquid scintillation counting has several advantages over ionisation counting. The major ones include:

1. Geiger counters are unable to detect soft emitters. However, significantly higher counting efficiencies are possible for low-energy beta emitters.
2. Ability to handle and count any type of sample—solids, liquids, suspensions, and gas. Different types of Geiger counters are needed for measuring various types of samples. Frequently, different kinds of accessories are required. Virtually any kind of sample (liquids, solids, suspensions, emulsions, gels, chromatograms, etc.) can be used in a scintillation counter and its radioactivity can be measured accurately.

Scintillation counting has several advantages over other detection methods, including a broad selection of scintillator materials, a quick response time, high detection efficacy, and a large detection area. In any radiation detection method, the counting efficiency (the ratio of the observed counting rate to the rate of radioactive disintegrations in the sample) is greatest when the maximum number of emitted particles reaches the detector and interacts with it in the desired way. Losses in efficiency result when the (isotropically emitted) particles escape in directions away from the detector or are stopped in it by some mechanism which does not result in a count. Beta rays lose their energy while passing through matter more rapidly than gamma rays of comparable energy and therefore have small ranges. There is absorption of low-energy beta particles while they are moving along their paths within and between the samples. In order to avoid such losses and to increase the efficiency, the distances between the sample and the detector must be decreased as much as possible. This is advantageous in liquid scintillation counting.

1. The ability to count different isotopes separately in the same sample: Geiger counters cannot differentiate between radiations of various isotopes and so cannot be employed to measure dual or triple labelled samples. These types of samples can be readily detected by a scintillation counter. Scintillation counters are highly automated.
2. Improvements in sample preparation techniques Sample preparation for scintillation counting is easier than that required for Geiger counters.

3. As fluorescence decay is very fast compared to the dead time in a Geiger-Muller counter, much higher count rates are possible using a scintillation counter.

3.5 Disadvantages

The instrument design has changed, which has largely solved the issues with scintillation counting. However, some of the disadvantages are:

(a) The high voltage applied to the photomultiplier produces electronic events in the system that do not depend on radioactivity but contribute to a high background count. This effect is termed "*photomultiplier noise*" and can be minimised by cooling the photomultiplier.
(b) *Quenching* is the greatest disadvantage. Quenching means a decrease in the efficiency of the energy transfer process in the scintillation solution. This occurs when the energy transfer process is obstructed at any point. There are three types of quenching. Dirty scintillation vials interfere with light that the sample emits before it reaches the photomultiplier, and cause *optical quenching*. *Colour quenching* occurs if the sample is coloured. This results in some of the light being absorbed before it leaves the sample vial. Less light reaches the photomultiplier. If a substance in the sample interfere with the transfer of energy from the solvent to the primary flour or from the primary flour to the secondary flour, the process is called *chemical quenching*. This is the most difficult form of quenching to correct. The extent of "quenching" varies considerably with different sample materials.
(c) The cost per sample of scintillation counting is considerably higher than Geiger counting.

Despite the drawbacks, the liquid scintillation method is the method of choice for assaying low energy β emitting samples, especially ^{14}C and ^{3}H. In addition, the relative simplicity of sample preparation, the adaptability of sample form, and the commercial availability of dependable counting systems with high-capacity automatic sample changers have made this detection method the most useful when analysing a large number and variety of samples.

Check Points

- State the mechanism of radiation detection by a solid scintillation counter.
- What is a scintillation cocktail?
- Explain the working of the liquid scintillation counter.
- Write about the advantages of scintillation counter over gas-filled detector.

Thus, a solid scintillation counter is a radiation detector which consists of a scintillation crystal to detect radiation and produce light pulses, while a liquid scintillation counter detects the scintillation produced in the scintillation cocktail by radiation. The advantages of a scintillation counter include its efficiency, high precision, and counting rates. Solid scintillators are safe and convenient alternatives to liquid scintillation cocktails.

4 Semiconductor (or Solid State Detector)

Key Concept
In semiconductor detectors, the fundamental information carriers are electron-hole pairs generated along the path of the charged particle (primary or secondary) through the detector.

These are the surface barrier detectors. It is not generally possible to measure the amount of ion pairs produced by ionising radiation in a solid, as is possible in a gas.

However, under certain conditions and for a class of substances known as semiconductors, it is possible to measure the amount of such ion pairs produced. They have both free electrons and free "positive holes", both of which contribute to the flow of current. In a semiconductor, both the "holes" and the electrons occupy energy levels that make them loosely bonded, and both charge carriers can be moved through the material by applying an electric field. When a distinct atom type is introduced (doped) into the primary semiconductor material, either holes or electrons become the predominant current carriers. Such materials are then referred to as p-type or n-type, respectively. Whether p- or n- type, the material is electrically neutral.

A substance with electrical conductivity that falls somewhere between that of a metal and an insulator is referred to as a *semiconductor*. Semiconductor materials are highly sensitive to γ radiation and are used to detect even very weak radiation of this type. Instead of the ion pairs which carry the charge in other types of detectors, electron–hole pairs are the cause of the electric current produced here. γ radiation causes the ejection of electrons from the valence bonds of the semiconductor, leaving a number of positive holes. The electrons travel towards the positive electrode while the positive charges (holes) travel in the opposite direction through successive exchanges of electrons between the neighbouring lattice sites. The number of electron-hole pairs formed is almost 170 times the number of ion-pairs formed in a gas ionisation chamber, and thus the current produced is much more than that of the other types of detectors. This enables the detection of even a very small amount of incoming radiation. In contrast to gas chambers, where 30 eV is required to create one ion pair, a germanium-lithium diode requires only 3 eV to create one electron–hole pair in the depletion region.

4.1 Types of Semiconductor Detectors

The two most common semiconductor detectors use germanium and silicon crystals doped with a small amount of Li, i.e., Ge (Li) detectors and Si (Li) detectors. Ge (Li) detectors are used for the detection of X- or γ-rays, whereas Si (Li) detectors are primarily used for the detection of corpuscular radiation. The main advantage of a Ge (Li) detector over a scintillation detector is its excellent energy resolution. Ge (Li) semiconductors are excellent detectors of gamma rays, with a resolution ten times higher than NaI (Tl) scintillometers. A typical Ge (Li) detector can easily yield 1% energy resolution for ^{137}Cs γ rays, compared with 10% using a NaI (Tl) detector. However, the main drawbacks of Ge (Li) detectors, i.e., their low sensitivity, the necessity of maintaining them at a low temperature of 77 K (room temperature is about 300 K), and their unavailability in larger sizes severely limit their application in nuclear medicine. Moreover, Ge (Li) semiconductors need to be cooled by liquid nitrogen and are hence cumbersome and not suitable as field instruments.

In addition to the above, there are instruments known as gamma ray spectrometers that can distinguish between various peaks and thus enable the identification of the source of radiation. The summary of various types of detectors is given in Table 4.

Table 4 Comparison of detectors used for measuring radioactive events

S. No.	Group	Types of detectors	Detection	Use	Characteristics
1	Gas filled detectors	Ionisation chambers	α, β, γ	Radiation survey	Accurate dose measurements, slow response, non-pulse type
2		Proportional counters BF3 and He-3 gases	α, β, n	α, β Contamination survey	Good α, β, n discriminations
3		Geiger counters	β, γ	Radiation survey	Reliable, inexpensive High dead time/ saturation
4	Scintillation counters	NaI (Tl)	γ	Nuclear spectroscopy Lab-use and survey	Higher efficiency and lower energy resolution Rapid response Expensive
5		ZnS	A	Survey	Detect only
6		Liquid scintillators	Low β-energies	Contamination C-14, H-3,	High efficiency, expensive
7	Semi-conductor detectors	Germanium detectors (Ge (Li), HPGe), and Si (Li)	X-ray γ	Nuclear spectroscopy Lab-use	Higher energy resolution Lower efficiency Need N_2 cooling
8		Silicon surface barrier detectors	Alpha and charged particles	Nuclear spectroscopy Lab-use	Good energy resolution

5 Other Detectors

Two other types of substances, thermoluminescent and photoluminescent are also used as radiation detectors. These substances do not produce light immediately after interacting with radiation, but they can preserve a portion of the radiation's energy for a long time. At a later time, the accumulated energy is released in the form of light by either heating (thermoluminescent) the irradiated substance or exposing it to laser light (photoluminescent). Thermoluminescent detectors are used in radiation monitoring, and photoluminescent detectors have not finding use in nuclear medicine but are finding uses in diagnostic radiology, particularly in digital radiology.

Thus, semiconductor detectors are radiation detectors that use a semiconductor material such as silicon or germanium crystal as a detecting medium. The average energy required to form an electron-hole pair is around ten times less than the one required for gas detectors and 100 times smaller than for a scintillation counter.

Check Points

- What are semiconductor detectors?
- Explain the working of semiconductor detectors.
- Write down the advantages and disadvantages of semiconductor detectors over other detectors.

In the end, it can be concluded that radiation detectors are used to detect high-energy particles produced from different sources. There are three types of detectors: gas-filled detectors, scintillators, and semiconductor detectors. All three types of detectors are involved in the detection and measurement of radiation based on different properties. All these have their own advantages and disadvantages and are chosen according to the type of particle and the type of information required.

Further Reading

Agranoff BW, Kirk A, Frey SB (1984) A regional metabolic contrast method for the study of brain pathology. Ann Neurol 1 S(suppl): S93-S97. https://doi.org/10.1002/ana.410150717.

Bisen PS (2014) Laboratory protocols in applied life sciences. CRC Press Taylor & Francis Group. International Standard Book Number-13: 978-1-4665-5315-6 (eBook - PDF).

Caprette DR (1995) Detection and Measurement of Radioactivity. Experimental Biosciences. http://www.ruf.rice.edu/bioslabs/methods/radioisotopes/rad1.html

Cherry SR, Sorenson JA, Phelps ME (2012) Physics in Nuclear Medicine. Elsevier Health Sciences.

Detecting Radiation in our radioactive world (2014) Teacher Resource Guide. American Nuclear Society.

Dirk Hünniger (2009) Geiger Mueller Counter with Circuit-en.svg

Domenech H (2017) Radiation Safety Management and Programs. Springer. ISBN: 978-3-319-42671-6.

Dorf RC (ed) (2000) The Electrical Engineering Handbook. ISBN: 0-8493-2177-8 ISSN: 1097-9409. CRC Press LLC

Enge HA (1966) Introduction to Nuclear Physics. Addison-Wesley Publishing Company. ISBN: 10: 0201018705

Feigelson RS. Scintillator Materials Group. Stanford University.

Gilmore GR (2008) Practical Gamma-Ray Spectrometry. John Wiley & Sons, Ltd. Print ISBN:9780470861967.OnlineISBN:9780470861981.https://doi.org/10.1002/9780470861981.

Girolamo BD, Nessi M (2014) 8.23 - Silicon Trackers. Reference Module in Biomedical Sciences Comprehensive Biomedical Physics: Science Direct, Elsevier 409-418. https://doi.org/10.1016/B978-0-444-53632-7.00628-6.

Heggie J, Liddell N, Maher K (2001) Applied Imaging Technology (4th edn.) ISBN 1-8752713-3-3 490 pages.

Kalender WA (2000) Computed Tomography: Fundamentals, System Technology, Image Quality, Applications. MCD Verlag, Wiley. ISBN: 978-3-895-78644-0

Kasprzak KS, Dencker L, Larsson BS, d'Argy R, Ullberg S, Tjalve H, Sigee DC, Randerath K, Randerath E, Ekins R, Barrio JR, Vis RD, Liu Y, Qu T, Wang X, Parak F (1991) Isotopic and nuclear analytical techniques in biological systems: A critical survey. Pure and Applied Chemistry 63 (9): 1269-1306. https://doi.org/10.1351/pac199163091269.

Khan NT (2017) Autoradiography: Detection and Analysis of Radioactive Entities. J Biom Biostat 8:4 https://doi.org/10.4172/2155-6180.1000361.

Knoll GF (2000) Radiation Detection and Measurement (3 rd edn.) John Wiley & Sons, Inc.

L'Annunziata MF, Kessler M (2012) Chapter 7 - Liquid Scintillation Analysis: Principles and Practice. Handbook of Radioactivity Analysis (3rd edn.). 423-573. Science Direct. Elsevier. https://doi.org/10.1016/B978-0-12-384873-4.00007-4.

McAlister JM (1979) Radionuclide Techniques in Medicine. Cambridge University Press. ISBN: 0521294746.

McParland BJ (2010) Nuclear Medicine Radiation Dosimetry: Advanced Theoretical Principles. ISBN: 978-1-84882-126-2. springer

Norman V, Herbert V (1958) Radioisotope Techniques in Clinical Research and Diagnosis. Butterworth London.

O'Reilly PH, Shields RA, Testa HJ (1986) Nuclear Medicine in Urology and Nephrology. Butterworths, ISBN: 0632059869.

O'Shea B (ed) (2010) Textbook of Psychological Medicine 5th edn. Published By Dr Brian O'Shea.

Parker RP, Smith PHS, Taylor DM (1984) Basic Science of Nuclear Medicine. 2nd edn. Churchill Livingstone. ISBN-10 0443024197

Photomultiplier Tubes. Basic and Applications 3rd End. (Edition 3a) (2007) Hamamatsu Photonics KK.

Poston JW (2003) Dosimetry. Encyclopedia of Physical Science and Technology (3rd edn.) 603-650. Science Direct, Academic Press. https://doi.org/10.1016/B0-12-227410-5/00185-X.

Radiation Safety for Radiation Workers. Chemical and Radiation Protection Safety Department (2005) University of Wisconsin - Madison.

Radioactivity measurements principles and practice (1988) International Journal of Radiation Applications and Instrumentation. Part A. Applied Radiation and Isotopes 39 (8): 717-937. Science Direct, Elsevier. https://doi.org/10.1016/0883-2889(88)90117-7

Ramón SJ, Xavier LX (2012) Handbook of instrumental techniques for materials, chemical and biosciences research. Centres Científics i Tecnològics UB (CCiTUB). ISBN: 978-84-615-5373-0.

Rinck P (2001) Magnetic Resonance in Medicine. Wiley Blackwell. 252 pages.

Seco J, Clasie B, Partridge M (2014) Review on the characteristics of radiation detectors for dosimetry and imaging. Physics in Medicine & Biology 59 (20): R303. https://doi.org/10.1088/0031-9155/59/20/R303.

Sharon M, Sharon M (2021) Nuclear Chemistry. 2nd end. Springer. ISBN: 978-3-030-62018-9.

Shenai S (1992) Optimized trench MOSFET technologies for power devices. IEEE Transactions on Electron Devices 39 (6): 1435 - 1443. https://doi.org/10.1109/16.137324

Sharon M, Sharon M (2021) Nuclear Chemistry (2nd edn.). Springer Jointly published with ANE Books Pvt. Ltd. ISBN 978-3-030-62017-2. ISBN 978-3-030-62018-9(eBook). https://doi.org/10.1007/978-3-030-62018-9.

Shimada M, Abe S, Imahayashi T (1982) Studies on regional blood flow of the mouse using whole- body autoradiography of 14C-iodoantipyrine. Histochemistry 75(2):179-90. https://doi.org/10.1007/BF00496009.

Solon EG (2012) Whole-Body Autoradiography and Microautoradiography in Drug Discovery and Development: Part IX. Bioanalytical Methods of Studying Metabolism. Encyclopaedia of Drug Metabolism and Interactions. https://doi.org/10.1002/9780470921920.edm103.

Sorokin YI (1999) Radioisotopic Methods in Hydrobiology. Springer. ISBN: 978-3-642-59934-7.

Talboys M (2016) Monte Carlo simulation applied to uncertainties in iodine-123 assay and thyroid uptake measurement. PhD Thesis. Cardiff University, School of Engineering.

Wilson K, Walker J (eds) Principles and Techniques of Biochemistry and Molecular Biology (6th edn.) Cambridge University Press

Use of Radiation in Diagnosis

Key Learning Objectives
- SPECT and its uses
- PET and its applications
- Characteristics of radiotracers
- Construction and working of a gamma camera
- Nuclear imaging and medical imaging techniques
- Diagnostic radiopharmaceuticals and their applications
- The importance of radiotracers in medical diagnostic techniques

Overview
In the last chapter, we discussed the methods used for the measurement of radiation. Radioisotopes are becoming more and more important in our lives. They are often used in medicine, industry, and scientific research, and the latest applications are continuously being developed. In this chapter, we will be dealing with the application of radiation in the diagnosis of various diseases.

Key Concept
A small amount of radiotracers introduced inside the patient gets distributed throughout the body and emits gamma rays, which are used to get more dynamic pictures of the organ or tissue.

A. Pathak, *Tools and Techniques in Radiation Biophysics*,
https://doi.org/10.1007/978-981-99-6086-6_8

Radiations form an essential part of medical diagnostic procedures because radioisotopes can be dispersed all over the body by its own metabolism; their position inside provides a concept of structure as well as the processes occurring within. From the pictures obtained, we can assess the damage caused by various disease conditions, and in some cases, these may be helpful in deciding the future course of treatment for that particular disease, as will be discussed in the succeeding chapters. Many of the instruments used for diagnostic purposes are available in the Department of Nuclear Medicine at various hospitals. In several cases, radioisotopes have no replacement, and in many of the applications, they are more efficient and cheaper than other alternative techniques.

1 Diagnostic Techniques

Key Concept
Radioactive iodine is used to understand the functioning of the thyroid gland.

In nuclear medicine, various diagnostic techniques are used that make use of radioactive tracers that release gamma rays from within the body. In addition to the available imaging devices that detect the gamma rays emitted from inside, these radiations are useful to study the metabolic and other physiological processes taking place in different body parts such as the thyroid, bones, heart, liver, and many other organs. These can be comfortably imaged, and their functional disorders can be revealed using different specific radioisotopes and imaging devices. Thousands of hospitals worldwide utilise radioisotopes, and around 90% of the applications are for diagnosis. Nuclear medicine techniques employ a negligible amount of radiation to obtain information about a body—the functioning of different organs, the ongoing biological processes, and the diseased state of a particular illness. In most cases, this data is used by clinicians to accurately diagnose the patient's illness.

Key Concept
A hot spot is the area which has a greater amount of radionuclides.
A cold spot is an area that does not absorb any radionuclide.

During the initial stages on the nuclear medicine field, radioactive iodine was used for measuring the activity of the thyroid, and this has got a huge success. The gland is unable to differentiate between radioactive and non-radioactive forms of iodine and so absorbs all iodine isotopes. On the basis of this absorption, the activity of the thyroid gland can be assessed. The radiolabelled drugs, for example, isotopes of iodine incorporated into thyroxine and other suitable formulations, are used to deactivate a hyperactive thyroid. This development has led to the thyroid gland being the most understood of all endocrine organs and its hormones being the best recognised of all endocrine secretions. This knowledge is fundamental to the treatment of a large number of patients because the incidence of disease involving the thyroid gland is very high, second only to diabetes mellitus among the endocrine disorders. Because the brain is a critical and delicate organ, methods for identifying brain tumour were undertaken in detail. In 1948, radioisotopes were used for detecting tumours.

Using radio-tagged molecules that were specifically absorbed by brain tumour, physicians were able to precisely detect and locate brain tumours, that allowed for improved diagnosis and more accurate surgery. After the success of brain scanning, scans of other organs like the liver, spleen, gastrointestinal system, gall bladder, lymphomas, and bones were also developed and successfully used for diagnosis purposes. The radiotracers used in nuclear scanning should possess the following characteristics:

1. They should have short half-lives and a daughter product that is non-toxic and non-radioactive. They decay and emit radiation while the diagnosis is being done, then decay so quickly that very little radiation is given off after the diagnosis so as to minimise the risk to patients.
2. They should emit penetrating gamma rays that can be detected by the detectors outside the body.
3. Their chemical properties should be such that they get concentrated in diseased areas (making a "*hot spot*") or cause the diseased areas to reject them (making a "*cold spot*").

Check Points

- What is the importance of radioisotopes in medical diagnosis?
- How is radio iodine helpful in assessing thyroid functioning?
- What are the properties of radiotracers?

Thus radiation is an indispensable component of medical diagnostic methods. The images obtained by the radiation emitted from within the body help in assessing the harm caused by the disease to a particular organ.

2 Diagnostic Radiopharmaceuticals

A radiopharmaceutical is a drug that contains a radioisotope bound to a biological molecule. These molecules take the radioisotope to particular organs, tissues, or cells. The radio-pharmaceuticals can be used in both diagnostic and therapeutic applications. For diagnostic studies, a radioactive isotope is introduced into the patient, and its concentration in a particular organ can be studied by taking either a two-dimensional picture or a three-dimensional picture. The tracers used in nuclear medicine are normally short-lived and are linked to chemical molecules that permit specific physiological activities to be investigated. These are given to the patient by injection, inhalation, or orally. There are a large number of molecules that are absorbed by particular tissues and organs. They are called targeting agents or molecules, for example, glucose. The brain consumes large quantities of glucose. Using this kind of knowledge, radiopharmacists or scientists can choose a suitable targeting molecule, tag it with a specific radioisotope, and use it as a radiotracer that concentrates in a particular tissue or organ of the body. After that, this radiopharmaceutical goes into the body and is metabolised and excreted in the usual manner. The number of radiopharmaceuticals to be used clinically is growing rapidly, thus permitting the medical fraternity better access to the detailed information about the properties of the various types of tumours.

Diagnostic radiopharmaceuticals have a number of applications. They are useful in examining blood flow to the brain, assessing the functioning of the liver, lungs, heart, and kidneys, assessing bone damage, etc. They also find use in sports medicine for diagnosing stress fractures that are not generally detected with X-rays.

For diagnostic work, radiopharmaceuticals are used in such minute quantities that it is just sufficient to get the desired information prior to the disintegration of a radiopharmaceutical that disintegrates and gets excreted out of the body. The radiation dose obtained by the patient is similar to the dose that is received from X-rays; however, it has the added advantage of being non-invasive and having the ability to detect organ functions. This makes it a preferred and powerful diagnostic tool. Typically, the radiation released by the radioisotope used for imaging disappears completely after 1 day because of radioactive decay and usual body excretion. The most common radioisotopes used for diagnostics and imaging in nuclear medicine use is ^{99m}Tc. Technetium has the ability to combine with a large number of molecules. The molecule to be labelled with ^{99m}Tc is chosen so that it is specific for the organ under investigation. This isotope, ^{99m}Tc, possesses a half-life of 6 hours and decays by emitting a gamma ray, forming the stable isotope ^{99}Tc.

Hospitals performing these procedures have ^{99m}Tc generators that contain the ^{99}Mo isotope that decays to ^{99m}Tc. This is then tagged with a specific vector molecule to produce a specific radiopharmaceutical that is administered (injected) into the body and is useful in scanning various organs like the brain, bone, liver, spleen, kidney, and lung cancer, and as well as for blood flow anomalies. When ^{99m}Tc is converted into ^{99}Tc, the emitted gamma radiation is measured by a gamma camera. The amount of radioactivity measured represents information on the functioning of the organ; high or low radioactivity detected by imaging with a gamma camera represents the hyperactivity or underactivity of the organ, respectively. It can also be used for diagnosing certain forms of cancers; for example, ^{99m}Tc-MDP (methylene diphosphonate) is mostly used to locate bone metastases.

Key Concept
^{99m}Tc is a radionuclide nuclear agent that is approved by the FDA for diagnostic purpose.

Other isotopes used in diagnosis include ^{123}I, ^{131}I, ^{18}F, ^{201}Tl, ^{111}In, and ^{67}Ga. The frequent diagnostic nuclear procedures performed are:

- Cardiac imaging (using ^{99m}Tc or ^{201}Tl),
- Bone scintigraphy, including tumour metastases (mainly ^{99m}Tc),
- Lung investigation (mainly ^{99m}Tc),
- Thyroid (^{99m}Tc or Iodine-^{123}I/^{131}I).
- Kidney function analysis (^{99m}Tc).
- Tumour staging (^{18}F-FDG).

Some of the most important ones are depicted in Table 1.

These different radioactive isotopes go to specific organs, emit gamma rays, and then they are imaged using different techniques.

Check Points
- What is a radiopharmaceutical?
- What are the properties of a radiopharmaceutical?
- What are target agents?
- Discuss the applications of diagnostic radiopharmaceuticals.
- How are radiopharmaceuticals better than diagnostic X-rays?

Table 1 Different radioactive isotopes and their uses

S.No.	Radioisotope	Applications
1	^{14}C	Metabolism
2	^{47}Ca	Investigating cellular function and bone formation
3	^{57}Co	Pernicious anaemia diagnosis
4	^{60}Co	Instrument sterilisation
5	^{51}Cr	Studies on RBC and blood volume determination
6	^{137}Cs	Treatment of cancer
7	^{67}Cu	Cancer treatment
8	^{3}H	Metabolism of drugs
9	^{125}I	Diagnosis of thyroid disorders
10	^{131}I	Treatment of thyroid disorders; diagnosis of kidney infection, blockage, and functional imbalance
11	^{32}P, ^{33}P	Molecular biology and genetic research
12	^{35}S	Genetics and molecular biology
13	^{75}Se	Protein studies
14	^{99m}Tc	Blood flow studies and organ imaging
15	^{201}Tl	Cardiac imaging
16	^{133}Xe	Lung ventilation and vascular flow

Therefore, the radiopharmaceuticals are used to produce images of a target tissue or organ by a process called scintigraphy and are used to investigate certain medical conditions or treat certain diseases. This technique is a potent diagnostic tool because of its non-invasiveness and its capacity to disclose the activity of the organ or tissue. Mainly, ^{99m}Tc is used in the production of radiopharmaceuticals. Apart from this, some other isotopes are also used for scanning specific organs.

3 Nuclear Imaging

Nuclear imaging is a procedure that uses gamma-emitting radioisotopes for creating an image. As already discussed in the last section, these radioisotopes are tagged with a particular molecule depending on the target organ. That molecule is called a radiopharmaceutical. To prepare a radiopharmaceutical of interest, a specific radioisotope is tagged with a molecule that concentrates in a specific organ or tissue. The radiopharmaceutical can be given to the patient orally, or is injected, or inhaled, and is then detected using a gamma camera, which creates a digital image.

Before the advent of nuclear imaging, mainly medical imaging techniques were preferred. Nuclear imaging differed significantly from other imaging techniques such as computerised tomography (CT), magnetic resonance imaging (MRI), or X-rays. Nuclear imaging is not able to provide high-resolution anatomical images of body structures, as is possible with other medical imaging methods. However, it measures the functioning of a particular body part either by measuring the blood flow or the distribution of the radioisotope in a particular organ.

For example, from a bone scan, one can identify the prevalence of secondary cancer "spread" up to 2 years ahead of a regular X-ray. It draws attention to the microscopic remodelling efforts that the skeleton is making in order to combat the invading cancer cells. Also in nuclear imaging, the radiation source is inside the body, and the results reveal the position and amount of radioisotope inside the patient's body. Very little uptake of the radioisotope seen as a "cold spot" in the image suggests that blood flow is restricted. However, a "hot spot" may suggest excess radioactive uptake in a particular tissue or organ that may be due to some disease. This nuclear imaging system can successfully image both bones and soft tissues.

SPECT/PET cameras are the special devices used in nuclear imaging, in which radiopharmaceuticals produce images of specific organs or tissues by a process called scintigraphy. We have already discussed that radionuclide imaging depends on the capability to detect radiation generated from the radiotracer that is taken up by a particular organ, and a digital image is generated by the scintillation camera or gamma camera. It produces images in a non-invasive manner, reflecting the functioning of the organ under investigation. The earliest method developed used single photons to view organs from many angles with a gamma camera. The camera generates an image of all the locations from where radiation is being emitted; this image is magnified and seen on a monitor for signs of abnormal conditions.

Key Concept
A gamma camera, also called a scintillation camera, is the most commonly used imaging device in nuclear medicine.

A scintillation camera, or gamma camera, contains a collimator, a scintillation crystal (NaI), photomultiplier tubes, a preamplifier, an amplifier, a pulse-height analyser, and recording and display devices.

Key Concept
A collimator is a fine processing layer of a gamma camera to receive photons for radioactive source. It enables the gamma camera to accurately localise the radionuclide in the patient's body.

The camera is operated by a built-in computer. The detector head (consisting of a collimator, scintillation crystal, PM tubes, and amplifiers) is supported by a structure known as a gantry, which allows the detector head to be moved to the appropriate position for imaging a patient (Fig. 1). The crystal is positioned in front of light detectors that convert the emerging light into an electrical signal. Gamma cameras map the function and processes of the body, while X-ray imaging provides information about the anatomy and structure.

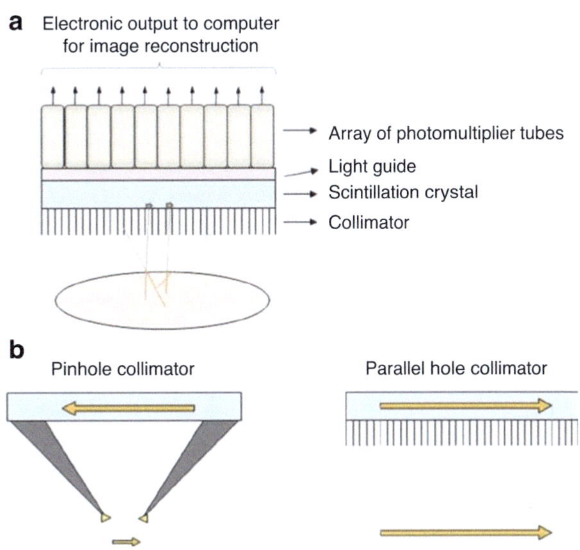

Fig. 1 Basic components of Gamma Camera. (**a**) Detector head (**b**) Different types of collimators Adapted from Radiolabelling Strategies for Radionuclide Imaging of Stem Cells. (*Adapted from: Esther Wolfs & Catherine M. Verfaillie & Koen Van Laere & Christophe M. Deroose. Stem Cell* Rev and Rep DOI 10.1007/s12015-014-9575-3)

Key Concept
Hal Anger is the inventor of the gamma camera used in nuclear medicine imaging.

Anger-type scintillation cameras are more frequently used in nuclear medicine (Fig. 2). It contains a large scintillation crystal that detects radiation from the whole field of vision simultaneously, allowing it to record both dynamic and static images of the target area in the patient.

Detectors have normally a large (about 50 cm diameter) circular or rectangular NaI(Tl) scintillation crystal with a thickness of 1 cm. In front of the scintillation crystal is a collimator to restrict the field of view and prevent the gamma radiation from outside the field to reach the crystal (Fig. 3). The collimator is normally a conical plate of lead with one to three holes at the top (pinhole) and is mainly used for imaging small organs, like a thyroid gland, tight joints, or infant kidneys. Mostly

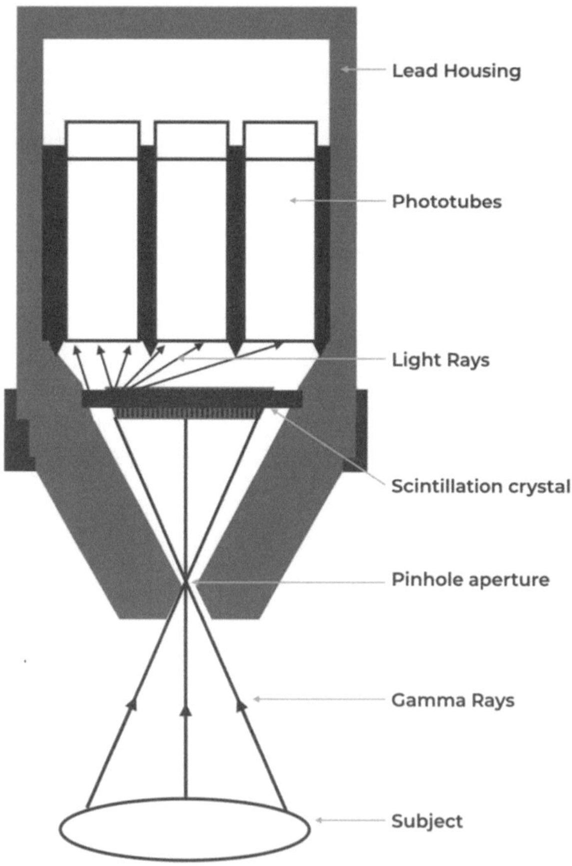

Fig. 2 Anger's Gamma camera

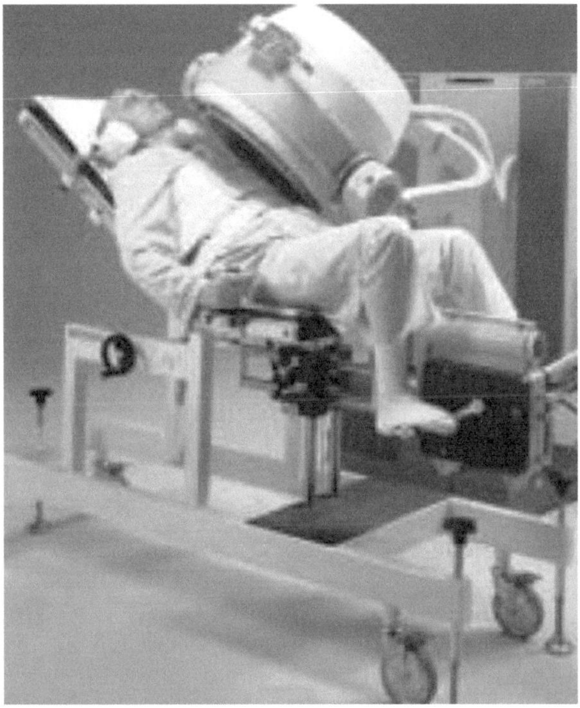

Fig. 3 Gamma camera used to detect dysfunctional activity in body organs. (*Adapted from U.S. Nuclear Regulatory Commission or U.S. NRC*)

parallel-hole collimators are used that have holes parallel to each other and perpendicular to the face of the detector. The collimator offers a very good resolution but poor sensitivity. Recently, collimators with converging holes (called fan-beams) have also been developed for small organ imaging, like the brain. But for imaging large organs, collimators with diverging holes are used. Gamma cameras have several photomultiplier tubes (maximum up to 90) mounted on the rear of the crystal with optical grease. These are used to generate and store images digitally for further processing. The spatial resolution of an image increases with the number of photomultiplier tubes. Over the years, many sophisticated improvements to the cameras have been made depending on the need; however, the basic operation principles remain the same.

As already discussed, the processing of the images is done by computers. The usage of digital computers in nuclear medicine has grown significantly, and all investigations are now analysed digitally. The basic function is to construct the correct image and to maintain camera execution parameters, like high voltage in the PM tubes and photo peak setting using the pulse height analyser, as high as possible.

The analogue-to-digital converter is required to transform gamma camera data into digital form. A square matrix with sizes ranging from 32 by 32 to 1024 by 1024 is used by the computer to represent the detector's surface area. Each element in this matrix is referred to as a pixel. Each pixel's value represents the total number of pulses recorded at that spot on the crystal. In this manner, the data is stored in frame mode. Thus, it is necessary to specify the matrix size, the number of images (frames) included in the study, and the duration of data collection per frame, or the total number of counts to be gathered per frame.

4 Emission Computed Tomography

According to the kind of radionuclide used, there are two types of emission computed tomography (ECT):

1. Single photon emission computed tomography (SPECT), that uses gamma-emitting radionuclides; and.
2. Positron emission tomography (PET), that uses positron-emitting radionuclides.

4.1 Single Photon Emission Computerised Tomography (SPECT)

This is currently the major scanning method to identify and track a variety of medical issues. A gamma- emitting radioisotope is delivered into the patient's bloodstream normally through injection. The radioisotope may be a simple dissolved ion (an isotope of gallium (III) or may be a specific ligand that is bound to a radioisotope to form a radioligand that binds to specific tissues and is seen by a gamma camera.

The SPECT cameras have the similar detector as planar cameras, but the gantry allows the detector to be rotated around the patient. At small angle increments, the minimum rotational angle is 180 degrees. The path around the patient can be circular or elliptical, and modern cameras offer the option of the so-called body contouring trajectory to reduce the distance between the detector's forehead and the patient's body surface. The data is stored digitally for additional processing and image reconstruction, that is done in the form of slices in three perpendicular planes, i.e., transverse, coronal (frontal), and sagittal directions. Then, this data can be changed to display fine slices of the body along any desired axis.

Therefore, SPECT images monitor the extent of biological activity in each area of the three-dimensional region analysed. The emission of radiation from the radionuclide indicates the rate of blood flow in the capillaries of the areas that were imaged.

There are different categories of SPECT cameras with varying numbers (one, two, or three) of detectors. The SPECT image is a static image, so a substantial image information density must be attained. Around 20–30 minutes are required to collect sufficient data, and images are reconstructed using mathematical algorithms (filtered back projection), which increase image characteristics and suppress artefacts. SPECT is being used in conjunction with gamma imaging studies, so that a true three-dimensional representation can be helpful, for example, in tumour imaging, infection (leucocyte) imaging, thyroid imaging or bone scintigraphy. It can also be used to provide details about the localised function of internal organs, such as cardiac and brain imaging.

4.2 Positron Emission Tomography (PET)

Key Concept
A positron is an antiparticle of an electron. It is also called an "anti-electron".

Positron emission tomography (PET) is another medical imaging procedure that is useful in revealing the biochemical or metabolic functions of tissues or organs.

PET makes use of radiotracers, a special camera, and a computer to help investigate organ and tissue functions. PET may be able to identify an early diagnosis of the disease before it is noticeable on other imaging tests by recognising cellular alterations. A positron-emitting radionuclide is administered, usually via injection, and accumulates in a specific tissue. The radionuclide emits a positron as it decays. This positron quickly integrates with nearby electrons and results in the simultaneous emission of two gamma rays in opposite directions. These two photons have the same energy of 511 keV but are emitted at an angle of 180 degrees and are detected by the two detectors connected in coincidence. A PET camera can find these and provide extremely precise information about their source.

In current PET systems, up to thousands of detectors are arranged in circular rings (up to 16) and are arranged in a sequential manner. These rings connect the opposite detectors with the help of a coincidence circuit. The breadth of the array of detectors determines the field of view. No collimator is required for this. The data is collected, and the image reconstruction is done by the same method as SPECT.

PET is a more precise and sophisticated technique using isotopes produced in a cyclotron, like ^{11}C, ^{18}F, ^{15}O, and ^{13}N. These can be easily incorporated into the biological tissue and, therefore, can be readily used as biotracers. PET is a clinical as well as a research tool. It is used extensively in clinical oncology (medical imaging of tumours and metastases) and for the clinical diagnosis of certain diffuse diseases, like different types of dementia. It is a crucial research instrument for mapping normal human brain and heart function and for assisting in the development of new drugs.

The study of human brain function is one of the most recent developments in positron emission tomography (PET) scanning. As glucose is the brain's principal fuel source, doctors can pinpoint the areas of the brain that are actively metabolising a glucose analogue by labelling it with ^{18}F and relate them to their function.

^{18}F is being used as the tracer because it has been shown to be the most accurate, non-invasive means of detecting and evaluating most malignancies. This can be done using fluorodeoxyglucose (FDG), which is an analogue of glucose. The concentration of tracer imaged will show tissue metabolic activity because it tracks glucose uptake in the area. The likelihood of cancer metastasis is investigated using this tracer. Other visible capabilities, like language, object identification, and fine motor coordination, have lately been linked to increased brain activity in particular regions. Therefore, this method has ushered in a new era of brain research.

A PET scan also evaluates important body functions, such as blood flow, oxygen use, and sugar (glucose) metabolism. It is also used in the prognosis of cancer therapy to determine blood flow to the heart muscle, determine the effects of a heart attack or myocardial infarction, identify areas of the heart muscle that would benefit from a procedure such as angioplasty or coronary artery bypass surgery, evaluate brain abnormalities such as tumours, memory disorders, seizures, and other central nervous system disorders, perform imaging of skeletal muscles, and map normal heart and brain function.

PET-CT, which combines PET with CT or MRI scans for co-registration of the two images, allows for a 30% better diagnosis than with a conventional gamma camera alone. In order to create a three-dimensional (3D) image of the patient's body, a CT scan takes a series of X- rays at various angles. PET/CT combines the structural information of a CT scan with the functional information of a PET scan, considerably assisting in correct staging, therapy response evaluation, and early recurrence detection. Images obtained from PET and CT scans taken together can be used to pinpoint the precise anatomical location of aberrant metabolic activity within the body. More accurate results can be obtained from the combined scans, as has been demonstrated one image for a more accurate diagnosis than the two scans performed separately.

Check Points

- What is nuclear imaging?
- How is nuclear imaging different from medical imaging techniques?
- Explain the construction of a gamma camera.
- Describe the mechanism of image formation in SPECT.
- How is SPECT different from PET?

In summary**,** a gamma camera contains a collimator, a crystal, and a number of photomultiplier tubes that are connected to a computer. It records every emission from the radio tracer in the body and converts it into an image. SPECT emits gamma radiation that is measured directly, while PET tracer positrons collide with electrons up to a few millimetres away, releasing two gamma rays in opposite directions.

Further Reading

Araoju T, Abayaid M, Rotten MJC, Misra S (2017) Segmentation and three-dimensional reconstruction of lesions using the automated breast volume scanner (ABVS). Int J Med Robotics Comput Assist Surg 13: e1767. Wiley Online Library. https://doi.org/10.1002/rcs.1767.

Carter CS, Dalley JW (eds) (2012) Brain Imaging in Behavioral Neuroscience. Part of the book series: Current Topics in Behavioral Neurosciences (CTBN, volume 11). Springer. ISBN: 978–3–642-28,711-4.

Chap. XI: Isotope Production and Cancer Therapy Accelerators (2014) Engines of Discovery 166–183. https://doi.org/10.1142/9789814417204_0011.

Harbin MH, Kasak A, Ostrem JD, Dengel DR (2017) Validation of a three-dimensional body scanner for body. composition measures. Eur J Clin Nutr 72(8):1191–1194. https://doi.org/10.1038/s41430-017-0046-1.

Heike U (1986) Single-photon emission computed tomography by inverting the attenuated Radon transform with least-squares collocation. Inverse Problems 2 (3): 307–330. IOP Publishing Ltd. https://doi.org/10.1088/0266-5611/2/3/008.

Keservani RK, Kesharwani RK, Sharma AK (2016) Chap. 10 - Nanobiomaterials involved in medical imaging technologies. Nanobiomaterials in Medical Imaging: Applications of Nanobiomaterials 8: 303–337. Science Direct. William Andrew. Applied Science Publishers. https://doi.org/10.1016/B978-0-323-41736-5.00010-8.

Moini J, Badolato C, Ahangari R (2020) Chap. 6 - Diagnostic Evaluation: Epidemiology of Endocrine Tumors. 109–132. Elsevier. https://doi.org/10.1016/B978-0-128-22187-7.00002-5.

Use of Radiation in Therapy

Learning Objectives

- Advantages of radioisotopes over X-rays
- Properties of radio isotopes to be used in nuclear medicine
- Use of diagnostic radiopharmaceuticals in therapy
- Therapeutic radiopharmaceuticals
- External beam therapy
- Internal radiation therapy
- Systemic radiation therapy

Overview

As discussed in the previous chapter, X-rays for diagnostic purposes, such as locating broken bones in the patient, were the most prevalent medical application of radiation. Later, radioisotopes opened a new door in the field of diagnosis. The natural tendency of certain organs to preferentially absorb specific radioisotopes, combined with the ever-improving detection techniques of nuclear medicine, made it possible to use radioisotopes to increase the contrast between various body parts. Due to the different densities of hard and soft tissues, X-rays can distinguish between them. Radioisotopes offer the added advantage of being able to distinguish between different types of tissues based on their metabolic function as opposed to their physical density alone.

A. Pathak, *Tools and Techniques in Radiation Biophysics*,
https://doi.org/10.1007/978-981-99-6086-6_9

1 Advantages of the Use of Radioisotopes

Radioisotopes offer two advantages for their use. First, extremely small amounts of radioisotopes are required, as very sensitive instruments have been developed nowadays to measure such a small amount of radiation. Secondly, there are specific radioisotopes for specific organs and tissues, and in this way, they can be targeted to a particular body part. For instance, radioactive sodium iodide behaves in a similar manner in the body as sodium iodide found in iodised salt.

The thyroid gland concentrates iodine, which is then converted into the hormone thyroxine. Likewise, other radioactive atoms can be routed to other organs such as bone marrow, red blood cells, the liver, the kidneys, or made to remain in the bloodstream, where they are measured with the aid of the appropriate instruments.

Key Concept
Radiation therapy uses high-energy particles or waves—X-rays, gamma rays, or photons—to destroy cancerous cells.

In a number of medical conditions, particularly cancers, the purpose of radiation therapy is to destroy damaged or cancerous cells using radioisotopes while affecting normal tissues as little as possible. They are also essential for non-curative or palliative care. For instance, they can improve the patient's quality of life and alleviate the intense pain caused by bone metastases, which can be very intense. Indeed, rapidly dividing cells are especially susceptible to radiation damage, allowing for the control or elimination of certain cancers through irradiation. Utilising external irradiation or internally sealed sources is a possibility.

The energy of the radiation is the most important factor in radioisotopes' usefulness. When radioisotopes are used as a treatment, their energy absorbed by the target cells or tissues is either used to destroy damaged tissue, especially malignant tissue, or to inhibit certain functions of these cells or tissues. Radiation doses that are properly calculated and administered to a patient can be used to produce the desired effects with minimal side effects. Although the amount of energy associated with a radiation dose is small, this energy is released in a manner that causes significant changes in the molecular composition of individual body cells.

Therefore radioisotopes are currently a preferred option for diagnosis as well as for therapy. Even in a very small amount, the radioisotopes can be targeted to a specific site where they cause the destruction of the cancerous cells while causing no harm to the neighbouring normal ones. Basically, the cells and tissues absorb the small amount of energy associated with radioisotopes to destroy the malignant cells, producing minimal effects on the surrounding healthy cells. They are also helpful in relieving the discomfort associated with the disease, for example, in cases of bone metastasis.

2 Therapeutic Radiopharmaceuticals

Key Concept
Therapeutic nuclear medicine is developing as an alternative choice of treatment in oncology and pain management.

Rapidly dividing cells are more sensitive to radiation and susceptible to damage. By irradiating a specific area, certain cancerous growths can be controlled or eliminated. This is termed radiotherapy. Nowadays, significant research is being conducted globally on the attachment of radioisotopes to highly specific biological compounds, resulting in the formation of radiopharmaceuticals. Radioisotopes act as sources of concentrated radiation and are concentrated mainly within the affected tissues or organs. The dose can be calculated to have the greatest possible therapeutic effect without harming surrounding healthy tissues. Several scientists are optimistic about the potential of these radioisotopes in the recovery or even cure of certain diseases by attaching these radiopharmaceuticals to a particular calculated radioactive radiation exposure.

2.1 Properties of Therapeutic Radiopharmaceuticals

Therapeutic radiopharmaceuticals have two main characteristics: physical and biochemical.

Key Concept
Diagnostic radiopharmaceuticals are used to determine the disease or disorder while therapeutic radiopharmaceuticals are used to treat a previously diagnosed disease or disorder.

- The *physical* properties include the physical half-life of the radionuclide, type of emissions occurring, energy of the radiation(s), daughter product(s) produced, method of production of radionuclides, and purity of the radionuclide. Therapeutic radionuclides have a physical half-life in the range between 6 h and 7 days. A very brief physical half-life is almost unfeasible and limits the delivery flexibility, whereas radionuclides with a long half-life will be retained with the patients and also irradiate the neighbourhood for a longer duration. A radiopharmaceutical should decay into a stable product like ^{153}Sm which ultimately decays into ^{153}Eu. Even if the daughter product is slightly unstable, it must be extremely short-lived and should degrade within a few hours into a stable state. Radioisotopes that emit short-range particles (α or β) are preferred for therapy. These types of particulate radiation produce very high ionisation per length of distance travelled. As a result of their power to lose all their energy in a very short distance, they are capable of inflicting significant local damage (such as the destruction of cells or palliative pain management for bone cancer or arthritis). Such isotopes remain in the body for a longer period than those used in imaging; this is desirable and is done intentionally in order to increase the efficiency of treatment; however, this remains limited to a few days only. The most common examples of therapeutic isotopes that have been used successfully to treat numerous benign and malignant conditions are ^{131}I, ^{32}P, ^{90}Sr, ^{90}Y, ^{188}Rh, ^{177}Lu, etc.

- The *biochemical* properties of radiopharmaceuticals include targeting the tissue, retention of radioactivity in the target tissue, *in vivo* stability, and toxicity. A particular tracer is selected depending on the target site, and it should provide selective concentrations and prolonged retention in the affected tissue or organ while exposing the normal tissue as little as possible. Moreover, irrespective of the mechanism of introduction of radionuclides into the patient's body, the ratio of the concentration of radiopharmaceuticals in the tumour to those in other organs or tissues should be as high as possible so that the radiation can be delivered optimally to the target tissue or organ while reducing the radiation dose delivered to the normal tissues to the minimum.

Key Concept

Delivery of radioactive atoms to tumour-associated targets is called radiopharmaceutical therapy.

Check Points

- What are therapeutic radiopharmaceuticals? How are they different from the diagnostic ones?
- Describe the characteristics of radiopharmaceuticals.
- What type of radiation is used for therapy?
- Give examples of therapeutic radionuclides.
- What is the mechanism of action of therapeutic radiopharmaceuticals in the target tissue or organ?

In summary, therapeutic nuclear medicine is nowadays emerging as a preferred choice of treatment in oncology as compared to its traditional use in endocrinology and rheumatology. The main reason is its non-invasiveness and relatively low toxicity compared to conventional chemotherapy. For these, radiopharmaceuticals are used, which are actually unique medicinal formations emitting radiation. Radiopharmaceuticals must possess particular characteristics in their composition and manufacture so that they can be successfully used to diagnose and treat health complications.

There are three main methods for providing radiation therapy:

3 External Radiation (or External Beam Radiation) Therapy or Teletherapy

External radiotherapy is the commonest form of radiation therapy that uses machinery to direct high-energy radiations from outside the body to the inside of diseased cancer cells and to shrink tumours, and therefore can treat extensive regions of the body. It is a type of local treatment, i.e., it is used to treat a specific targeted area of the body. The beam of radiation can deliver a high dose of radiation without affecting the surrounding healthy tissue. ^{60}Co has been utilised effectively to irradiate and treat cancerous tumours. ^{60}Co forms the core of an external irradiator, and the gamma radiation emitted from the source is focused on the tumour of the patient. Although ^{60}Co irradiators have been largely replaced by linear accelerators in recent years, AEC sponsorship was instrumental in their development, and they were responsible for numerous advances in radiation therapy.

Key Concept
Tomotherapy is a form of IMRT that employs a device that combines the capabilities of a CT scanner and the external- beam radiation machine.

3.1 Types of Particles Used

The three main types of particles used for external therapy are photons, protons, and electrons.

3.1.1 X-Rays or Photons

These are produced by a machine known as a linear accelerator (also called a linac). The radiation beam for X-ray or photon radiation therapy is generated by this machine. This is the most widely used radiation therapy machine. Special computer software adjusts the size and shape of the beam of radiation so that only the target tumour is irradiated while avoiding healthy tissue close to the tumour cells. X-rays are not used here as they are associated with low doses. Photons, on the other hand, possess a higher amount of radiation. Therefore, they can reach even the deep-seated tissues, and because of their higher energy, they move past the affected tissue and therefore also irradiate or expose the surrounding healthy tissue.

3.1.2 Protons

These are positively charged particles that are also capable of delivering higher doses to the target tissue. However, they offer an advantage over the photons in that they stop at the target tissue only, so the surrounding tissue remains unexposed to the radiation. A machine that is used to speed up protons is called a synchrotron or cyclotron. The high cost of these machines limits the use of proton beam therapy.

3.1.3 Electrons

These are negatively charged particles. They are not able to travel to deep-seated tissues. Therefore, they are preferred only for treating skin cancer and tumours that are near to the skin surface.

The external radiation therapy is given for several weeks to the patients during their out- patient visits to a hospital or treatment centre.

3.2 Types of External Radiation Therapy

External- beam radiation therapy is of different types:

3.2.1 Three-Dimensional Conformational therapy (3-D-CRT)

3-D-CRT is one of the most common types of external radiation therapy. This therapy uses the images obtained from CT, MRI, and PET scans to obtain detailed three-dimensional images of the treatment area. This allows physicians to design the treatment plan to aim the radiation at the target area more precisely so that higher doses of radiation can be used safely while minimising damage to healthy tissue.

3.2.2 Intensity-Modulated Radiation Therapy (IMRT)

This is a more advanced type of 3-D CRT. IMRT also targets the tumour from several directions and avoids healthy tissue. However, the performance is better than conventional 3-D-CRT because here smaller beams are used and the strength of the beams can be changed so as to deliver higher doses to a particular area of the tumour.

Key Concept
Simulation is the use of CT, MRI, and PET scans to plan the treatment area.

3.2.3 Proton-Beam Therapy

This treatment uses protons instead of X-rays because they have greater energy. The protons travel to the targeted tumour, deposit a precise dose of radiation, and are therefore able to destroy the cancer cells. In contrast to X-ray beams, the radiation is limited to the targeted tumour. This reduces the risk of harm to nearby healthy tissue. This therapy is relatively new, and expensive equipment is required. Consequently, it is not available at all medical facilities.

3.2.4 Image-Guided Radiation Therapy (IGRT)

IGRT is similar to IGMT. The only difference is that this type of radiation therapy allows for taking photographs of a patient throughout the course of treatment, i.e., before the beginning of the treatment and also during radiation therapy treatment. Repeated CT, MRI, and PET scans are taken to monitor the progress of treatment, i.e., the size and location of the tumour. The radiation dose can then be adjusted by comparing these images to the initial images that are to be delivered to the patient. This adjustment increases the efficacy of the treatment.

3.2.5 Stereostatic Radiation Therapy

Key Concept
GammaKnife is a type of stereo static radio surgery.

Stereostatic radiation therapy uses high-energy focused beams to treat a very small area of the tumour. This is mainly practised when the tumour cannot be safely handled with surgery. This treatment targets a large, targeted area of the tumour. This therapy is primarily used to treat brain and central nervous system tumours with well-defined borders.

3.2.6 Fast Neutron Therapy

Fast neutron therapy is a highly effective and unique form of external beam radiation therapy. It is frequently used to treat specific malignancies that are radioresistant or exceedingly challenging to eradicate with standard X-ray radiation therapy. A neutron can stop the growth of a tumour by destroying a tumour cell through a nuclear process that splits the atoms in malignant cells into pairs of distinct atoms that cannot rejoin.

Check Points

- What is photon radiation therapy?
- What is the aim of performing external radiation therapy?
- Name the machines used in external radiation therapy.
- Describe different types of external radiation therapy.

In summary, external beam radiation therapy uses a machine placed outside the body to direct radiation beams at cancer to destroy and shrink it. This therapy may also be used to ease the symptoms of patients in advanced stages of cancer. Mainly photons are used in most types of radiation therapy treatments, but tiny particles—electron and proton—may also be used in a few selected cases. Of these, electrons are used only for superficial cancers. Thus, external-beam radiation therapy can be given by different techniques and different types of radiations. Of these, the most common type is three-dimensional conformal radiation therapy.

4 Internal Radiation Therapy

In internal radiotherapy, the radioisotope element alone or attached to the biological compound (i.e., in the form of a radiopharmaceutical) is administered into the patient's body. This radiopharmaceutical concentrates at the diseased site and generates the prescriber's dose of radiation from that affected organ. Here, the radioactive source is always encapsulated so that it does not move to other parts of the body. Implants may also be used to place a radioactive source within the body or in close proximity to the tumour. These implants may be temporary or permanent and may require hospitalisation.

Key Concept

Internal radiation therapy is also called radiation brachytherapy.

4.1 Temporary Brachytherapy

Here the radioactive material is placed inside the catheter for a specific period of time, and then it is removed. In place of catheter, needles, wires, ribbons, tubes or special applicators can also be used. After these temporary implants are properly positioned at the desired site, the radiation source is inserted using a computer-controlled remote after-loading device; the process is called "after-loading". The radiation may be given at a low dose rate or a high dose rate, depending on the severity of the disease.

4.2 High-Dose-Rate (HDR) Brachytherapy

The applicator is loaded with a potent radioactive source, a patient is put in the applicator, and the patient is treated for several minutes. The source is removed after a few minutes, depending on the severity of the disease. This may be repeated twice per day for several days or once per day for several weeks. The radioactive substance is eliminated from the patient's body. Depending on the circumstances, the applicator can be left in place between treatments or inserted prior to each application. Sometimes the patient may occasionally need to be hospitalised if multiple-day treatments are given and the applicator is left in place.

Key Concept
High-dose rate brachytherapy is frequently performed as an out-patient procedure.

4.3 Low-Dose-Rate Brachytherapy

In this type of therapy, lower doses of radiation emanate from the implants over a longer period of time. Some implants are left inside for one to several days before being removed. The patient must remain in the hospital, occasionally in a specialised room, during treatment to prevent others from being exposed to radiation. If larger implants are used, then the patient has to lie still in bed to prevent the implant from moving. If smaller implants (like seeds or pellets) are used, they are left in place and never taken out. After a particular period of time, depending on the radioisotope used, the seeds stop emitting radiation. Because these seeds or pellets are roughly the size of rice grains, they rarely cause any issues if they are left inside a patient's body.

Key Concept
Seed is a non-radioactive metallic capsule in which the radioactive material to be used for internal radiation therapy is enclosed.

4.4 *Permanent Brachytherapy*

In this technique, the implants, which are approximately rice-grain-sized radioactive seeds, are permanently placed in or close to the tumour using a delivery device (either ultrasound or X-ray imaging) to ensure their correct placement. They deliver the vast majority of radiation around the implant site. The seeds lose their radioactivity over time. The implant(s) are no longer functional when the radiation has subsided. There is no need to remove them because they often remain *in situ* and pose no threat.

The type of radioisotope used (iodine, palladium, cesium, or iridium) depends on the type of treatment. For instance, ^{131}I is used in internal radiotherapy for the treatment of thyroid cancer, hyperthyroidism (an overactive thyroid), and some other rare forms of cancer like neuroblastoma (a cancer that arises from nerve cells), probably the most effective type of cancer treatment. It is mostly given in the form of capsules or mixed in a drink, but it is also injected into a vein. During treatment, the patient may need hospitalisation.

Implants containing ^{192}Ir are used especially in both brain and breast cancers. They are produced in wire form and are introduced through a catheter to the head or breast. The implant is not removed for the particular time after which it is removed and kept in shielded storage. This procedure is much cheaper than that using external radiation to provide therapy and also gives less overall radiation exposure to the body.

^{125}I is used for treating prostate cancer. ^{125}I seeds at low dosage rates are implanted in the prostate and remain there forever in the patient.

^{89}Sr as well as ^{153}Sm are used for treating a few forms of bone tumours. They help reduce pain and improve quality of life. ^{18}Re is a newer product for this. Quadramet is a commercially known radiopharmaceutical that incorporates ^{153}Sm and is used internally to reduce the pain that arises due to secondary bone cancers associated with breast, prostate, etc.

^{223}Ra is another radioisotope used to treat secondary bone cancer due to prostrate cancer that is no longer responding to hormonal therapy.

4.5 *Intracavity Radiation Therapy*

During this type of radiation therapy, the source containing radioactivity is placed inside a body cavity (space), like the rectum or uterus, using a tube or cylinder that fits in a particular body opening. The brachytherapy device may be placed at the required position manually or using computerised imaging equipment that includes either a CT scanner or an ultrasound machine so that the device is placed in an appropriate location accurately.

4.6 *Interstitial Radiation Therapy*

With interstitial radiation therapy, the implants are inserted into or near the tumour location, such as within the breast or prostate, but not in a body cavity. For this purpose, wires, balloons, and small seeds the size of rice grains are used. CT scans, ultrasounds, etc. may be utilised to direct the devices into their specific locations.

4.7 *Episcleral Radiation Therapy*

Here, the radiation source is attached to the eye. Mainly, melanoma of the eye is treated with this type of radiation therapy.

Internal radiotherapy uses higher localised dose of radiation to treat tumours as compared to that given in external radiation therapy. This has the advantage that the surrounding normal tissue is not irradiated. Internal radiotherapy is useful in the treatment of certain systemic cancers, for example, bone metastases, where it is impossible to use whole-body irradiation using external beam radiotherapy.

Check Points

- How is internal radiation therapy different from external radiation therapy?
- State the differences between temporary and permanent brachytherapy.
- What is the difference between low- and high-dose radiation therapy?
- Enlist the radioisotopes used in internal radiation therapy.

In summary, internal radiation therapy is mainly used to treat prostrate, breast, head, neck, and other malignancies. Here, the physicians insert microscopic radiation pellets into or around the tumour. Radiation released by the pellets kills cancer cells, sparing healthy tissue and organs. Radiation helps in the destruction or damage of the DNA composition of cancer cells. Damaged cancer cells cannot grow, multiply, and eventually die off. The procedure for brachytherapy may vary depending on the type and intensity of the disease.

5 Systemic Radiation Therapy

In systemic radiation therapy, the drugs containing radionuclides are given orally or injected intravenously. These drugs then travel through the blood throughout the body, locating and killing tumour cells. Systemic radiation therapy is used in the treatment of certain forms of cancer, like thyroid, bone, and prostrate cancer, or to relieve the patient of pain when the cancer has metastasised to the bones. These liquid radioactive drugs are called radiopharmaceuticals or radionuclides containing radioactive isotopes. Systemic radiation therapy is also called unsealed intensity radiation therapy, radioisotope therapy, radionuclide therapy, or radiopharmaceutical therapy. There are certain radioactive drugs that are also used to diagnose other health problems. Urine, saliva, and sweat will be radioactive in this case; therefore, those in close proximity to the patient must take the precautions advised by the medical staff.

Key Concept
Systemic radiation therapy helps in the treatment of cancer that targets the entire body.

The radioactive isotopes used for systemic radiation therapy include ^{131}I, ^{89}Sr, ^{32}P, and ^{223}Ra.

For example, radioactive iodine therapy (RAI, using ^{131}I) is given for thyroid cancer. After thyroid surgery, radioactive iodine is frequently administered to eliminate any remaining cancerous tissue. Radioactive iodine (^{131}I) has been used for decades to help treat selected thyroid cancer patients after surgery and for the treatment of hyperthyroidism, or Graves' disease. After the radioactive iodine pill is taken orally, ^{131}I moves through the blood and is taken up by the thyroid cells. Therefore, it is possible to have systemic radiation therapy with few side effects on the rest of the body.

Another example of a treatment used to possibly enhance outcomes for males with bone-only metastases is ^{223}Ra. This therapy is used when hormone therapies are no longer effective. It is an alternative that may be taken into account either before or after chemotherapy, although the two are not currently provided in tandem. Currently, a radiation oncologist administers ^{223}Ra via a series of six outpatient injections.

5.1 Types of Systemic Radiation Therapy

5.1.1 Radioimmunotherapy

Radioimmunotherapy uses a radiopharmaceutical called monoclonal antibodies that is combined with a small amount of radioactive material. The radioactive substance serves as a tracer that can locate and attach to the specific markers on the surface of the cancer cells; the monoclonal antibody is then delivered directly to the cancer cells. These antibodies are prepared and designed in such a way that they are able to attach themselves directly to the cancer cell and, therefore, damage the cells with very small amounts of radiation. There is less impact on the normal tissue that surrounds the affected tissue as a result of the employment of these unique antibodies.

Ibritumomab (Zevalin), which is used to treat certain lymphomas, is one instance of radioimmunotherapy.

Key Concept
Radiosensitisers are the substances that help radiation therapy better destroy tumours, e.g., fluorouracil (5- FU, Adrucil).

5.1.2 Peptide Receptor Radionuclide Therapy (PRRT)

In this procedure, a radiopeptide is created by mixing radioactive material having a unique protein known as a peptide. When administered, the radiopeptide seeks out and binds to specific cancer cells before directly bombarding the cells with a high dosage of radiation.

Key Concept
Radioprotectors are substances that protect healthy tissues near the treatment area, e.g., Anifostine (Ethyol).

5.1.3 Novel Targeted Therapies

These therapies are more specific in inhibiting the growth of cancer cells and may improve the efficacy of other treatments. Some targeted therapies, for example, employ drugs that inhibit the growth of new blood vessels that would nourish the cancer in order to prevent its growth. Other targeted therapies function by inhibiting the action of molecules called growth factors on the surface of cancer cells.

5.2 *Intraoperative Radiation Therapy*

Intraoperative radiation therapy refers to radiation treatment administered during surgery. In order to apply radiation directly to the tumour during surgery, the surgeon temporarily pulls healthy organs out of the way or covers them. Intraoperative radiation can be given either as external beam therapy or as brachytherapy.

5.2.1 Targeted Alpha Therapy (TAT)

This is a comparatively new field, especially for the control of dispersed types of cancer. In targetted alpha therapy, an α emitting radionuclide is attached to a biological molecule having targetting capabilities, for example, a monoclonal antibody

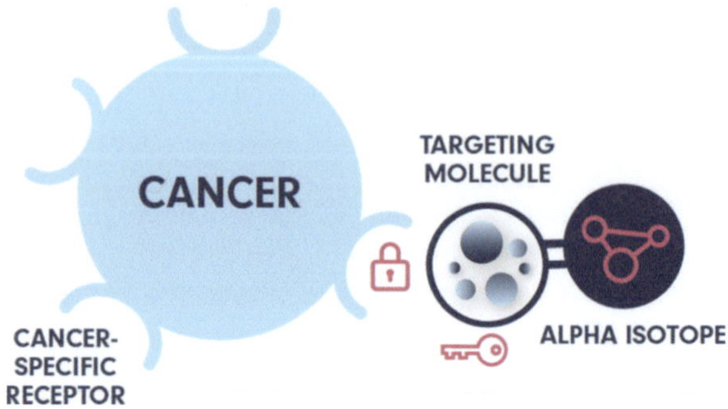

Fig. 1 Targetted alpha therapy. *(Adapted from https://*www.ionetix.com/wp-content/uploads/2021/09/IONETIX_WebsiteGraphics_Rebrand_Molecule.png. *Peter* Burke. Ionetix)

(mAb), with the aid of a bi-specific chelating agent (Fig. 1). Short-range alpha particle radiation restricts its cytotoxic effect to cancerous lesions and the surrounding tumour microenvironment while limiting its toxicity to non-cancerous cells.

Key Concept
Targeted alpha therapy represents an emerging treatment approach and provides the opportunity to bypass mechanisms of acquired resistance in selected tumours.

Theoretically, complex double-stranded DNA breakage brought on by the emission of high-energy short-range alpha particles may result in cell death. Compared to beta emissions, alpha emissions have a higher level of radiobiological efficacy, which means it takes fewer particle tracks to cause cell death. Clinically effective isotopes that emit alpha particles for cancer therapy should have a short half-life, which will limit long-term radiation exposure and allow for their production, preparation, and administration in clinical settings. ^{223}Ra dichloride is the first-of-its-kind, commercially available alpha therapy approved for the treatment of malignant melanoma, Given the established overall improvement in overall survival conferred by radium (^{223}Ra) for patients with metastatic castration-resistant prostate cancer, a number of other targeted alpha therapies are being evaluated in clinical trials across a wide range of tumour types.

The only commercially available targeted alpha therapy using ^{223}Ra dichloride has been approved for the treatment of patients with metastatic, castration-resistant prostate cancer who have bone metastases. ^{223}Ra has been proven to improve overall survival for patients with metastatic castration-resistant prostate cancer, and other targeted alpha treatments are currently being tested in clinical studies for a variety of tumour types. Targeted alpha therapy has so far shown its proof of concept in the clinical effectiveness of ^{223}Ra, and further research could result in further treatment choices for various cancer types. The clinical endorsement of $^{223}RaCl_2$ in 2013 was a significant step towards the broad adoption of TAT as a secure and efficient method of cancer treatment. Another key development in clinical use is the benefit of

^{225}Ac-prostate-specific membrane antigen therapy in metastatic castrate-resistant prostate cancer patients who are resistant to traditional therapies. The compounds designated ^{227}Th and ^{225}Ac seem to be good candidates and could be the new generation of drugs that can increase patient survival in a variety of therapeutic settings.

5.2.2 Boron Neutron Capture Therapy (BNCT)

This therapy is an innovative therapeutic approach aimed at improving the therapeutic responses for a number of difficult-to-treat tumours by traditional methods. This type of therapy uses boronated agents to deliver boron-10 preferentially to tumours, that yields ^{7}Li and an alpha particle after undergoing irradiation with neutrons (Fig. 2). The alpha particles have a short range and therefore specifically affect the tumour tissues, while more distal normal tissues remain unaffected. A boron-containing substance is injected into a blood vessel; the boron collects in tumour cells, and the patient is subsequently given radiation therapy using neutrons. Boron and neutrons interact to destroy cancer cells, without harming healthy cells. This therapy has been evaluated clinically for a wide range of disease sites, including glioblastoma multiforme, meningioma, head and neck cancers, lung cancers, breast cancers, hepatocellular carcinoma, sarcomas, cutaneous malignancies,

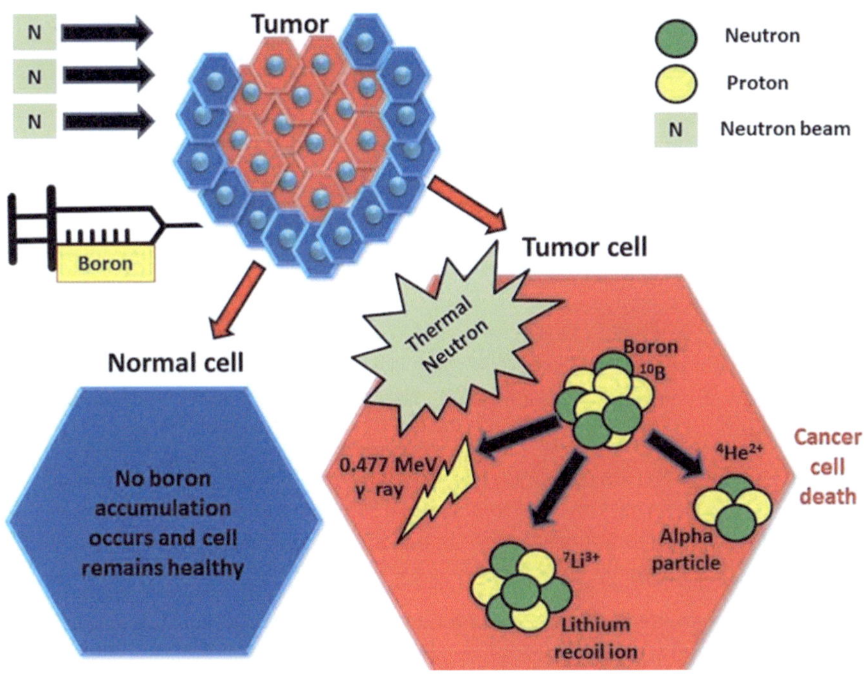

Fig. 2 Boron neutron capture therapy. (Adapted from Front. Oncol. Volume 11–2021. Article 521,820 by Malouff et al.)

extramammary Paget's disease, recurrent cancers, paediatric cancers, and metastatic disease. Neutron capture therapy with ^{10}B, which focuses on malignant brain tumours, is an experimental development of this. The patient is irradiated with thermal neutrons that are strongly absorbed by boron, resulting in the production of high-energy alpha particles that eradicate the cancer. Instead of bringing radioisotopes to the patient, the patient will be transported to a nuclear reactor. With any therapeutic procedure, the aim is to confine the radiation to well-defined target values for the patient.

The type of cancer, the size and location of the tumour, as well as the patient's general condition, will all influence the kind of radiation a doctor would advise.

Check Points

- Under what circumstances is systemic radiation therapy used?
- What types of radiopharmaceutical drugs are used in systemic radiation therapy?
- Give examples of the radioactive isotopes used in systemic radiation therapy.
- Write a note on targeted alpha therapy.
- Explain the mechanism of action of boron neutron capture therapy.

6 Objectives of Radiation Therapy

Different treatment objectives might be achieved with radiation therapy:

- To improve the efficacy of surgery,
- To help fight against the spread of cancer,
- To ease the signs and symptoms of advanced cancer,
- To cure the cancer, radiation therapy may be used to:
  - Remove cancers that haven't migrated to other body parts.
  - Lower the risk of the disease coming back following treatment with chemotherapy or surgery by eliminating small quantities of cancer that may still be present.
- To slow down the cancer as much as possible.
- To decrease the symptoms due to developing tumours and enhance the quality of life. This type of radiation therapy is called palliative care or palliation. Here, radiation therapy may be used to:
  - Reduce tumours that are affecting the quality of life, such as lung cancer that is associated with shortness of breath.
  - Reduce the size of tumour to relieve pain.

In summary, systemic radiation therapy may be used to treat thyroid, bone, and prostrate cancer. Here, radioactive drugs, or radiopharmaceuticals, taken orally or injected, are used to deliver the radiation to the target cells. These days, the application of clinical radiotherapy as a partner to immune therapy is opening a new field of investigation. In a variety of clinical settings, the clinical applications of targeted alpha therapy with various radionuclides and combination therapies are expanding. Boron capture neutron therapy is an emerging targeted therapy with promising outcomes and tolerable toxicity, according to preliminary clinical studies.

Further Reading

Ambrogio MW, Toro-González M, Keever TJ, McKnight TE, Davern SM (2020) Poly(lactic-co-glycolic acid) nanoparticles as delivery systems for the improved administration of radiotherapeutic anticancer agents . ACS Appl Nano Mater. 3 (11): 10565–10570. https://doi.org/10.1021/acsanm.0c02350.

Getting Internal Radiation Therapy (Brachytherapy) (2019) cancer.org | 1.800.227.2345. American Cancer Society.

Kaski JC (2004) Pathophysiology and management of patients with chest pain and normal coronary arteriograms (Cardiac Syndrome X). Circulation 109:568–572. https://doi.org/10.1161/01.CIR.0000116601.58103.62.

Malouff TD, Seneviratne DS, Ebner DK, Stross WC, Waddle MR, Waddle MR, Daniel MC, Krishnan S (2021). Boron Neutron Capture Therapy: A Review of Clinical Applications. Front Malouff TD, Trifiletti DM (eds) (2022) Principles and Practice of Particle Therapy. John Wiley & Sons Ltd. Print ISBN:9781119707516. Online ISBN:9781119707530. doi:10.1002/9781119707530. Oncol 11:601820. doi: 10.3389/fonc.2021.601820. eCollection 2021.

Parker C, Lewington V, Shore N, Kratochil C, Levy M, Park J, Saad F (2018) Targeted alpha therapy, an emerging class of cancer agents: A Review. JAMA Oncol 4(12):1765-1772. https://doi.org/10.1001/jamaoncol.2018.4044.

Simbawa E, Al-Johani N, Al-Tuwairqi S (2020) Modeling the spatiotemporal dynamics of oncolytic viruses and radiotherapy as a treatment for cancer. Hindawi Computational and Mathematical Methods in Medicine 2020 (3642654).10 pages. https://doi.org/10.1155/2020/3642654.

Suominen MI, Wilson T, Kakonen S, Scholz A (2019) The mode-of-action of targeted alpha therapy Radium-223 as an enabler for novel combinations to treat patients with bone metastasis. International Journal of Molecular Sciences 20(16):3899. doi: https://doi.org/10.3390/ijms20163899.

Wickstroem K, Hagemann UB, Kristian A, Ellingsen C, Sommer A, Ellinger-Ziegelbauer H, Wirnitzer U, Hagelin E-M, Larsen A, Smeets R, Bjerke RM, Karlsson J, Ryan OB, Wengner AM, Linden L, Mumberg D, Cuthbertson AS (2019) Preclinical combination studies of an FGFR2 targeted Thorium-227 conjugate and the ATR inhibitor BAY 1895344. Int J Radiation Oncol Biol Phys 105 (2): 410e422 Elsevier Inc. https://doi.org/10.1016/j.ijrobp.2019.06.2508.

Wu T-J, Chiu H-Y, Yu J, Cautela MP, Sarmento B, Neves J, Catala C, Pazos-Perez N, Guerrini L, Alvarez-Puebla RA, Vranješ-Đurić SJ, Ignjatović NL (2018) Nanotechnologies for early diagnosis, in situ disease monitoring , and prevention. Nanotechnologies in Preventive and Regenerative Medicine. 1-92. https://doi.org/10.1016/B978-0-323-48063-5.00001-0.

Use of Radiation in Cancer Therapy

Key Learning Objectives

- Radiotherapy
- Need for radiation therapy
- External beam radiation therapy
- Internal radiation therapy
- Mechanism of radiation therapy
- Imaging techniques

Overview

These days, radiation has become a valuable tool in medicine, research, and industry. In medicine, these radiations are used to diagnose illnesses, and higher doses are also used to treat diseases like cancer. Apart from the treatment, these days several sophisticated machines are available that help in imaging and hence diagnosing a particular disease and helps in planning the treatment of cancer patients by pin-pointing the areas of the tumour so that the surrounding normal tissues are exposed to very little radiation

1 Radiation Therapy

It has already been discussed in the previous chapters that radiation is present everywhere in the environment. All living beings are constantly exposed to different levels of radiation. Common factors that contribute to elevated risk such as those caused by nuclear accidents, natural background radiation, medical operations and

A. Pathak, *Tools and Techniques in Radiation Biophysics*,
https://doi.org/10.1007/978-981-99-6086-6_10

treatments, occupational exposures, etc. Radiation has a significant impact in the diagnosis of diseases and in therapy. Radiation therapy is now extensively used to cure cancer. This type of therapy is termed radiotherapy. Here, stronger dosages of radiation are used to destroy cancer cells and limit their growth by causing DNA damage. Radiation can cause cancer in most parts of the body, across all species, and at all ages. Radiation-induced leukaemia develops in 2–10 years, while solid tumours take 10–15 years and up to 40 years. Children and adolescents are twice as likely as adults to develop radiation-induced leukaemia, and the impact of perinatal radiation exposure is tenfold greater. The potential of radiation to cure cancer has been recognised since the early radiation studies; efforts to develop radiation therapy for cancer, however, increased after World War II. As it is already known, the cells usually multiply and divide to create new cells. However, cancer cells grow and divide faster than healthy cells. Radiation therefore acts by causing small DNA breaks within cells. These breaks inhibit the growth and division of cancer cells, causing them to perish. Therefore, radiation therapy causes DNA damage and also causes cells to cease growing or be destroyed. Radiation can also affect nearby normal cells, but the vast majority recover and resume normal function. This indicates that radioisotopes play a significant role in the treatment of disease, particularly cancer. Cancer is still thought to be a group of diseases with numerous potential origins. Chemicals being developed to treat cancer are advancing greatly. Radiation therapy is typically a local treatment, in contrast to chemotherapy, which typically exposes the whole body to anti-cancer medicines. In the majority of instances, it only targets and affects the area of the body being treated. Radiation and surgery remain the most effective methods for treating cancer, and there are numerous conditions for which radiation can provide relief. In addition, the innovative use of radioisotopes provides radiation therapy with a great deal more flexibility. Even as progress proceeds, this is anticipated to hold true for a number of years. Radiation treatment is planned for nearby healthy cells. Not only can radioisotopes be used to treat cancer, but other diseases as well. The use of radioisotopes to manage hyperthyroidism is possibly the most prevalent example. ^{131}I therapy was recognised as an effective alternative to surgery for the management of thyroid cancer at both the primary and metastatic sites. It shows how radioactive iodine was first used to monitor thyroid activity and then later to treat an abnormality in thyroid activity (hyperthyroidism). Radioactivity is also used in radiotherapy for carcinoma, with radium and cobalt therapies being the most common ones. Presently, long, thin needles of radioactive elements are drawn and inserted around the cancerous lesion to kill it. Drugs have been formulated that are preferentially "swallowed" and accumulated by cancerous cells, and thus cancer is selectively exposed to radioactivity and gets retarded or killed.

2 Goals of Radiation Therapy

Over 50% of cancer patients undergo radiation therapy. At times, the only necessary cancer treatment is radiation therapy.

Radiation therapy can be used to cure many different kinds of cancer, either on its own or in conjunction with other therapies. Radiation therapy is useless for treating cancer that has spread to multiple organs because it does not reach every part of the body. However, there are many stages at which radiation therapy is useful. These are mentioned below:

- **To cure early stage cancer:** Radiation is particularly effective against several types of cancer. Radiation might be used to make cancer shrink or completely go away, i.e., some tumours can be controlled by radiation therapy by shrinking them or preventing them from spreading. Radiation can also be used before surgery to reduce the size of the tumour (pre-operative therapy or neo-adjuvant therapy) or after surgery to prevent the cancer from returning (called adjuvant therapy). Radiation therapy, either alone or in combination with other interventions, has a high success rate in curing several types of cancer. Radiation is the preferred treatment over surgery because radiation may not do as much harm and the organ may function more normally following therapy.
- **To stop cancer from coming back (recurring) somewhere else**: From the original site of cancer, it might spread to different organs and tissues. Radiation therapy can therefore be used to eliminate any cancer cells before they develop into tumours. Some forms of lung cancer are particularly prone to spreading to the brain, so patients with these cancers may get preventative radiation therapy to the head.
- **To treat symptoms brought on by advanced cancer**: In certain cases, cancer has spread too far for treatment to be effective. However, some of these tumours can still be treated to reduce their size, allowing the patient to feel better and alleviating symptoms such as pain, difficulty swallowing or inhaling, and bowel obstructions. This is known as palliative radiation.

Key Concept
One of the first Americans to use X-ray radiation to treat cancer was a Chicago chemist and homeopathic physician named Emil Grubbe.

To eradicate cancer cells, radiation therapy employs high-energy gamma rays or other particles. These high-energy particles emit several types of radiation as they disintegrate. When radioisotopes are deposited in bodily tissues, the radiation they release can destroy cells within their range. If the cells that are exposed are healthy, this could be dangerous for the person. If the exposed cells are aberrant, however, the same procedure might be advantageous (for cancer cells, for example). Ionising radiation exposure is connected with an increased risk of acquiring cancer, particularly leukaemia, in the future. Although radiotherapy is currently used to treat cancer, it does not kill cancer cells right away. After receiving treatment for a few days or weeks, the cancer cells' DNA is damaged enough to cause them to die. Radiation therapy, while effective in reducing the number of cancer cells, does not completely eradicate the disease.

3 Different Kinds of Radiation Therapy

- External beam radiation therapy: The radiation is given to the body by a machine.
- Internal radiation therapy, also called brachytherapy: Radioactive particles deposited in the body close to the cancer cells could be the source of the radiation.
- Systemic radiation therapy: It employs radioactive materials such as radioactive iodine which travel through the circulation to eliminate cancer cells. About half of all cancer patients receive radiation therapy at some point during their treatment.

3.1 External Beam Radiation Therapy

This is the most prevalent form of radiation therapy. It delivers high-energy radiation from a machine outside the body, and it can treat large areas of the body if needed. It is a local treatment, meaning it focusses on a specific area of the body.

> **Key Concept**
> External beam radiation therapy uses a machine placed outside the body to deliver radiation, while internal beam radiation therapy uses implants or drugs to deliver radiation inside the body.

The radiation beam used in X-ray or photon radiation therapy is produced by a machine called a linear accelerator, or linac. The size and structure of the beam can be changed using specialised computer software. This helps target the tumour while avoiding healthy tissue close to cancer cells.

Photons, protons, and electrons are the three categories of particles used in external radiation therapy.

Photon beams: Photon beams are used in the majority of radiation therapy devices. These beams can reach deep-seated malignancies. A small amount of radiation is scattered along the path of the photon beams. However, once these beams reach the tumour, they continue on to normal tissue that is located past it. A device known as a linear accelerator generates photons.

Proton beam: Proton beams are also capable of reaching tumours located deep within the body, but unlike X-rays, they do not continue onward after they have reached the tumour. Proton beams are produced by special machines called particle accelerators, e.g., cyclotrons or synchrotrons. However, the use of proton beams in radiation therapy is limited due to their high cost.

Electron beam: The range of an electron beam is quite limited through body tissues. Therefore, their use is restricted to tumours on the epidermis or near the body's surface. Electron beams are released from the linear accelerator or a particle accelerator.

3.1.1 Types of External-Beam Radiation Therapy

External beam radiation therapy comes in a variety of forms, all of which have the same goal: to deliver the maximum prescribed dose to the tumour while sparing the normal tissue. A part of this type of therapy has already been discussed in the previous chapter.

- **3-Dimensional conformal radiation therapy (3D-CRT):** In order to plan treatment, detailed 3-D images of the malignancy are typically created using computed tomography (CT) or magnetic resonance imaging (MRI) scans. This process is called simulation. These images are analysed by the computer programme so that the precise shape of the tumour is known and the treatment is planned so that the radiation beam is only in the area of the tumour. As a result, the likelihood of adverse effects is reduced.
- **Intensity modulated radiation therapy (IMRT):** This is a more sophisticated form of 3D-CRT. Here, the radiation beam is administered to the tumour from multiple directions. In contrast to conventional 3D-CRT, which employs the same strength in each beam, IMRT varies the radiation intensity within each beam. This is done so that higher doses can be given to certain parts of the tumour. IMRT is superior to 3D-CRT in that it targets the malignancy while avoiding healthy tissue.
- **Proton beam therapy:** Instead of X-rays, protons are used in this treatment. A proton is a particle with a positive charge These high-energy protons have the ability to kill cancer cells. The protons travel to the targeted tumour and deliver the prescribed radiation therapy dose. The radiation does not penetrate the malignancy, unlike X-ray beams. This significantly reduces damage to nearby healthy tissue. This therapy is relatively novel and requires specialised equipment.
- **Image-guided radiation therapy (IGRT):** This method of treatment allows the physician to obtain images of a patient during treatment. i.e., for treatment planning prior to radiation therapy and during radiation therapy sessions. These images can then be compared to have better targetting of the tumour, i.e., improve the accuracy of the treatment and hence lessen damage to vital tissue. This also

helps to identify alterations in the tumour's size and location. This also allows the position and radiation dose to be adjusted during treatment if required.

- **Tomotherapy:** A device that combines external beam radiation equipment and a computed tomography scanner is used for this form of therapy. It takes the images just before treatment sessions so that the tumour is precisely targeted and no molecular tissue is saved. During treatment, it rotates around the patient and delivers radiation in a spiral pattern. However, this type of therapy lacks clinical trials, so it has limited use.
- **Stereostatic radiosurgery:** Small, well-defined lesions in the central nervous system are treated with focused, high-energy beams. It is preferred as an option if the tumour cannot be safely reached with surgery or if surgery is risky due to some other health issues. For example, using a gamma knife. Here, many small beams of radiation are aimed at the tumour from different directions in such a way that they have little effect on the tissue through which they pass and precisely deliver the radiation dose to the site.
- **Stereostatic body radiation therapy:** It is very similar to stereosoatic radiosurgery. The only distinction is that it is used for tiny, isolated tumours external to the brain and spinal cord, frequently in the liver, lungs, prostate, and kidney.

In all these types, in order to maximise therapeutic efficacy and minimise damage to healthy cells, radiation is administered in a series of treatments spaced at regular intervals.

3.2 *Internal Radiation Therapy*

This form of radiation therapy involves inserting needles, seeds, wires, or catheters with radioactive material into or close to a tumour. This form of radiation treatment is also known as brachytherapy, implant radiation therapy, or radiation brachytherapy. In the malignancy or the tissue around it, radioactive material is injected. Depending on the type of implant chosen, a hospital stay may be necessary. Internal radiation therapy comes in a variety of forms.

3.2.1 Brachytherapy (or Radioactive Implant Treatment)

An implant contains the radioactive material, which is referred to as a source. The implant may consist of seeds, wires, or discs. A minute amount of radioactive material is injected into the body, directly into or very close to the cancer or the area where it was removed. They deliver the majority of radiation to the implant area. Some radiation, however, may escape from the patient's body. To protect others from radiation exposure, this necessitates safety measures. The implant is left inside the body for a variable amount of time, such as several minutes, a few days, or perpetually. The implants lose their radioactivity over time, and the dormant seeds stay in the body. Brachytherapy is used to treat cancers such as prostate cancer, cervical cancer, and uterine cancer.

3.2.2 Radioactive Liquid Treatment

In this type of treatment, a patient has to undergo a scan after taking a radioactive liquid or tablet. A liquid can be consumed as a beverage, capsule, or injection. Examples of radioactive liquids along with their uses are shown in Table 10.1. This will help in precisely locating the area that needs treatment. The radioactive liquid gets collected in the area of cancer and has very little effect on the healthy cells of the body.

3.2.3 Intraoperative Radiation Therapy (IORT)

During surgery, an external beam or internal radiation therapy is used to administer radiation dose to the tumour. IORT enables surgeons to remove viable tissue beforehand. This treatment is effective when the tumour is close to vital organs. This form of therapy is administered to breast cancer patients who would ordinarily undergo a lumpectomy followed by radiation therapy. This method delivers radiation in one short session, allowing patients to complete treatment significantly faster than with standard radiation therapy.

3.2.4 Systemic Radiation Therapy

Patients are given radioactive chemicals that specifically target cancer cells through ingestion or injection. The radioactive material travels through the entire body rather than being aimed at one area. The radioactive substances exit the body through saliva, perspiration, and urine. These liquids have radioactivity. People in close proximity to the patient should therefore take precautions. Systemic radiation treatment uses radioactive medications (known as radiopharmaceuticals or radionuclides) to treat cancers such as thyroid, bone, and prostate cancer. These are radioactive substances in liquid form. In certain instances, a radioactive medication may be used to assist in the detection of cancers, such as bone metastasis (i.e., when cancer has spread to the bone).

Table 10.1 List of radioactive liquids

Radioisotope	Application
^{131}I	For both malignant and benign thyroid disorders
^{32}P	For some blood diseases
^{223}Ra	Blood cancer treatment (secondary bone cancer)
^{89}Sr	For secondary bone malignancies

3.2.5 Radioimmunotherapy

Systemic therapy is used in this case. Specifically, a little amount of radioactive material is transported to cancer cells using monoclonal antibodies, a special drug. This therapy delivers lower radiation dosages directly to the tumour. Non-cancerous cells are unaffected.

3.2.6 Radio Sensitisers and Radio Protectors

Radio sensitisers are substances that assist radiation therapy in more effectively destroying malignancies. Radioprotectors are chemicals that shield healthy tissues around the treatment site from radiation. Flurouracil (5-FU, Adrucil) and cisplatin (Platinol) are two radiosensitisers. Amifostine (ethyol) works as a radioprotectant.

3.2.7 Peptide Receptor Radionuclide Therapy (PRRT)

A radio peptide is created during this procedure by fusing radioactive material with a certain type of protein called a peptide. Once this radiopeptide locates and binds to cancer cells, a large dose of radiation is administered directly to them.

Radiation therapy may be the primary treatment or an adjunctive therapy. Before surgery, adjuvant radiation therapy can be used to decrease the size of the tumour, and afterward, it can be used to halt the growth of any remaining cancer cells. It is occasionally combined with chemotherapy.

4 Mechanism of Radiation Therapy

Key Concept
The four R's of radiobiology—repair of DNA damage, redistribution of cells in the cell cycle, repopulation, and reoxygenation of hypoxic tumour areas—determine the success or failure of normal therapeutic radiation treatment.

The primary principle of radiation therapy is to transport the highest possible dose to the tumour while delivering the lowest possible dose to the surrounding normal structures, thereby achieving a higher healing rate with tolerable morbidity

The cells normally proliferate and divide to form new cells. Cancer cells, on the other hand, develop and divide at a faster rate than normal cells. Radiation damages the genes (DNA) in cells. Cancer cells are unable to develop and divide when their genes are damaged. The cells eventually perish. Radiation therapy can either directly harm DNA or generate free radicals within the cells that can then harm

DNA. Radiation also causes microscopic DNA breaks inside cells. These breaks inhibit the growth and division of cancer cells, causing them to perish. However, during this process, the division of normal cells is also affected, resulting in a number of undesirable side effects. The length of time it takes for a cell to perish depends on the type and quantity of radiation. Tissues that grow rapidly, such as the epidermis, bone marrow, and intestinal lining, are frequently susceptible. Nerve, breast, brain, and bone marrow tissue, however, exhibit effects later. Radiation therapy is typically a local treatment, as opposed to chemotherapy, which typically exposes the entire body to cancer-fighting drugs. In the majority of instances, it only targets and affects the area of the body being treated.

When radiation deposits enough energy in organ tissue to cause ionisation, this tends to break molecular bonds, and modify the molecular structure of the irradiated molecules. Actually, the ionising radiation penetrating cells strike molecules at random. Radiation can decompose molecules into highly reactive free radicals, which can then cause additional chemical damage. This can kill cells or alter their genes to stop them from developing. Some of this direct and indirect damage will ultimately affect chromosomes and epigenetic factors that regulate gene expression. Some of this damage will be repaired by cellular mechanisms, but some of the remedies will be flawed, and some chromosome abnormalities will be permanent.

There are two types of ionising radiations with different energies:

- Photon radiation (X-rays and gamma rays).
- Particle radiation (electrons, protons, neutrons, alpha particles, beta particles, and carbon ions).

The more the energy, the deeper the radiation can penetrate tissue. The high energy of photons passing through the body breaks DNA bonds and prevents cell replication. Electron, proton, and neutron particle beams can also be used to irradiate areas of the body and disrupt the replication of malignant cells. Their mechanism of action has already been covered in the preceding section.

In addition, alpha and beta particles can be used to treat cancer, though they are more commonly used in medical imaging. Carbon ions produce radiation that is used to treat malignancies that do not respond to conventional radiotherapy. This is a heavier particle, though. Consequently, depending on where it ends its path, it can cause more damage to cancer cells and surrounding tissue than other forms of radiation.

Radiation damage can result in a stable, partially functional cell that can proliferate and developing into cancer, especially if tumour suppressor genes are disrupted. According to recent study, mutations do not arise immediately after irradiation. Instead, it appeared that surviving cells had acquired genetic instability, resulting in a higher mutation rate in later generations. After years of incubation, the cell will proceed through several stages of neoplastic transformation, culminating in the creation of a tumour. Cell morphological alterations, acquisition of cellular immortality (loss of normal life-limiting cell regulatory systems), and modifications that encourage cancer growth are the three major separate stages of neoplastic transformation.

Decades may pass between radiation exposure and a cancer diagnosis. Those cancers that may advance as a result of radiation exposure are identical to those that develop spontaneously as a result of exposure to other carcinogens.

5 Imaging

Imaging is essential for the planning and administration of radiotherapy. Nuclear medicine employs radioactive material within the body to examine organ or tissue function (for diagnosis) or to target and eliminate damaged organs or tissues (for treatment). Imaging with ionising radiation is restricted to the use of X-ray beams in diagnostic radiology and gamma rays in nuclear medicine, while radiation oncology uses a much wider range of ionising radiation, including neutrons, protons, and heavier charged particles. In nuclear medicine, the radiation that leaves the patient is of interest, but in diagnostic radiology, the radiation beam that passes through the patient is of interest. However, in the field of radiation oncology, what matters is the amount of energy that is deposited in the patient by a radiation source that is either outside the patient (external beam radiotherapy) or inside the tumour itself (brachytherapy). The screening process for a number of tumour forms now includes the use of cutting-edge diagnostic imaging techniques like computed tomography (CT), high-resolution magnetic resonance imaging (MRI), and ^{18}F-flourodeoxyglucose positron emission tomography (FDG-PET).

Positron emission tomography (PET) scans use radionuclides that release positrons. Nuclear medicine imaging includes PET scans. It is a non-invasive imaging method that delivers a functional view of organs and tissues, as opposed to CT scanning and MRI scanning, which only provide anatomical images. A CT scan uses X-rays to show a cross-section of the body. The images are more clear than standard X-rays. MRI uses a large, powerful magnet, and radio waves to create a similar picture. With the help of PET, doctors can examine and diagnose patients from a functional and biochemical standpoint. That is to say, it is able to detect metabolism within the tissues of the body.

During the 1970s and 1980s, PET was mostly employed in academic settings. In the early 1990s, the use of PET in hospitals and diagnostic clinics rose as medical communities realised its importance in clinical applications, particularly in oncology for cancer staging, evaluating treatment regimens, and monitoring the effects of therapy using appropriate radiotracers.

Since PET's spatial resolution is inferior to that of CT and MRI, anatomical and functional images from CT and PET must be fused. Software techniques are used to accomplish the fusion. Since the introduction of the first commercial PET/CT scanner into clinical practice in 2001, combined PET/CT has proven to be a significant advancement for the detection of primary tumours, distant metastases, recurrence after treatment, staging, restaging, and even monitoring therapy response in the majority of cancers. The positron-emitting radionuclides are affixed to clinically useful biological markers that are used in studies involving various metabolic processes and are therefore used in cancer diagnosis and treatment. PET assesses the function of tissues and organs using radioactive materials called radiotracers and radiopharmaceuticals, a specialised camera, and a computer. These radiopharmaceuticals bind to particular proteins in the body. Fluorine-based radiotracers are the most frequently used radiotracers in PET. The radiopharmaceutical fluorodeoxyglucose (FDG), which is the most frequently used tracer in studies involving glucose metabolism in cancer diagnosis, is composed of ^{18}F radionuclide affixed to the biological marker deoxyglucose. This molecule resembles glucose. As cancer cells are metabolically more active, they may absorb glucose more readily, i.e., at a higher rate. The emission of gamma rays from a radio tracer is detected by a specialised camera, and the attached computer generates images. PET scans show this higher prevalence.

Consequently, PET provides information on the metabolic function of organs or tissues by detecting how cells metabolise certain compounds, such as glucose. PET detects malignant cells with a high degree of sensitivity by detecting an increase in radiolabelled glucose metabolism, even at an early stage when other imaging methods may miss them. A PET scan may also be performed to determine the stage (i.e., how far the cancer has spread and if it is present in other organs and tissues); to determine if cancer treatment is effective; to determine if cancer has returned (recurred) after treatment or spread to other locations; and to diagnose nervous system and cardiovascular diseases. The high susceptibility of radioisotopes and the unique biocharacter of radiotracers, which are molecularly targetted radiopharmaceuticals, contribute to PET's prospective high sensitivity and specificity.

In a PET investigation, a positron-emitting radionuclide is administered via injection or inhalation. The radionuclide travels through the circulatory system to a specific organ. The nucleus emits a positron that travels a brief distance, up to a few millimetres, to collide with an electron in tissue and be destroyed by an available electron. This process typically results in the emission of two gamma photons, each with an energy of 0.511 MeV, moving away from the point of production in virtually opposite directions. The resulting gamma rays are detected by the PET system and converted into images. The radioactive signature of the radiopharmaceutical in the body is picked up by the detectors. A computer analyses the patterns and creates three-dimensional colour images of the scanned area. On a PET image, various hues or intensities of luminosity represent various degrees of tissue or organ function.

The radionuclides used in PET studies are created by bombarding a suitable stable radionuclide with protons from a cyclotron, resulting in the production of positron-emitting radionuclides that are then affixed to clinically useful biological markers. Other commonly used positron-emitting radionuclides besides ^{18}F include ^{11}C, ^{13}N, ^{15}O, ^{82}Rb, ^{64}Cu, and ^{124}I.

In clinical research environments with on-site cyclotrons, ^{11}C is used. FDG is the most frequently used oncologic PET tracer and the only one designated by the Food and Drug Administration (FDA) for routine clinical use among ^{18}F or ^{11}C-labelled radiotracers used in oncologic PET applications. ^{13}N-ammonia is preferred for PET myocardial perfusion due to its superior imaging characteristics, i.e., it allows for the sensitive and specific detection of coronary heart disease. However, its physical half-life is only 10 min, necessitating onsite cyclotron production. PET is used to study oxygen metabolism, cerebral blood flow, and blood volume in the heart, brain, and testicles using a simple chemical form of ^{15}O (half-life 2.05 min).

FDG is used in more than 90% of oncologic PET imaging procedures. Due to the increased metabolism of glucose by tumour cells, the uptake of FDG is significantly enhanced in the majority of cancer types, including lung, colorectal, oesophageal, stomach, head and neck, cervical, ovarian, and breast cancers, as well as melanoma and the majority of lymphomas.

Other metabolic PET imaging agents, such as radiolabelled amino acids, amino acid analogues, and choline, have also demonstrated favourable results in imaging brain tumours, prostate cancer, and other tumours; however, compared to FDG, far less research has been conducted on these oncological imaging agents. In addition to PET tracers based on increased metabolism, other tracers have been developed based on other characteristics of tumour biology and the tumour microenvironment, such as proliferation, hypoxia, angiogenesis, and apoptosis. As a proliferation indicator, ^{18}F-FLT has shown promising outcomes in evaluating tumour response to therapy and predicting early response during treatment. As a result, ^{18}F-FLT has the potential to become the next FDA-approved PET radiotracer.

Few PET tracers that successfully target particular cancer biomarkers exist. With the discovery of oncologic biomarkers, the evolution of novel PET tracers with high precision for targeting oncologic biomarkers will become an essential area of research. Current oncologic therapy has evolved from cytotoxic treatment to personalised therapy targetting specific molecular biomarkers, which can be assisted by PET imaging. With the progress of highly specific molecular probes, PET will play a significant and integral role in the diagnosis and monitoring of cancer and other diseases.

By selectively concentrating boron molecules in tumour cells and exposing the tumour cells to epithermal neutron beam radiation, boron capture neutron therapy is another technique that is emerging as a promising cancer treatment. Neutrons react with boron to destroy cancerous cells without harming healthy cells. Neutron capture therapy is a binary system that employs two distinct components to produce a therapeutic effect. Each component individually is not tumoricidal, but in combination, they are fatal to cancer cells. BNCT refers to the nuclear reaction that happens when ^{10}B, a stable isotope, is irradiated with low-energy thermal neutrons to produce alpha particles (^{4}He) and recoiling lithium nuclei.

$$^{10}B + n \rightarrow Li + {}^{4}He \tag{1}$$

The result is a high linear energy transfer (LET) alpha particle. Along their extremely short path (10 m), these electrons deliver high energy. As a result, high-energy deposition is restricted to a single cell's diameter. Following thermal neutron irradiation, only neoplastic cells containing ^{10}B are destroyed. Targeting is predominantly accomplished by precisely placing boron drugs within the tumour rather than by aiming the beam.

The most elusive characteristic of BNCT is its ability to deposit an enormous dose gradient between tumour cells and normal cells. This justifies its clinical application in the treatment of malignant cells, sparing normal, healthy cells. It is a non-invasive therapeutic procedure for treating locally spreading malignant tumours such as primary brain tumours and recurrent head and neck cancer. This is a two-step procedure: first, the patient is injected with a tumour-localising drug that contains the non-radioactive isotope ^{10}B, which has a high propensity or cross-section to capture slow neutrons. In the second stage, the patient is exposed to epithermal neutrons, either from a nuclear reactor or, more recently, from an accelerator. After losing energy as they penetrate tissue, neutrons are absorbed by the capture agent, which then emits high-energy charged particles that can specifically destroy tumour cells that have absorbed sufficient quantities of ^{10}B.

Check Points

- What are the objectives of radiation therapy?
- What is external beam therapy? Explain its different types.
- Describe the types of internal radiation therapy.
- Explain the mechanism of radiation therapy.
- Write a note on PET.
- What is the Boron Neutron Capture Theory?

Despite its adverse effects, radiotherapy can be highly effective in cancer treatment. Forty percent of cancer cures incorporate the use of radiotherapy as part of the treatment plan, either as a single modality or in conjunction with other therapies like surgery and chemotherapy. However, it takes some time for the beneficial effects to become apparent. Thus, radiotherapy may be used in the early phases of cancer as well as after the disease has begun to spread. It can be used to thoroughly cure the cancer (curative radiotherapy), thereby enhancing the benefits of other treatments like chemotherapy or surgery (neo-adjuvant radiotherapy). A number of advances in imaging have occurred in the recent past to precisely deliver the radiation to the affected tissue only, minimising the radiation exposure to the surrounding normal tissues. So patient outcomes are improved.

Further Reading

Bandazhevsky YI (2003) Chronic Cs-137 incorporation in children's organs". Swiss Med Wkly 133 (35–36): 488–90.

Brenner DJ, Hall EJ; Hall (2007). Computed tomography--an increasing source of radiation exposure". N Engl J Med 357 (22): 2277–84. https://doi.org/10.1056/NEJMra072149.

Eriksson D, Stigbrand T (2010) Radiation-induced cell death mechanisms. Tumour Biol 31(4): 363- 72. https://doi.org/10.1007/s13277-010-0042-8.

Giannouli S (2006) Computer-Assisted Radiation Therapy (CART). Wiley Online Library. https://doi.org/10.1002/9780471740360.ebs1433.

Kaur K (2021) Throwback: The evolution of treatment methodologies over the decades in breast cancer care. Times Now Digital.

Kale VG, Vandana B (2018) Exploring the average information parameters over lung cancer for analysis and diagnosis. Computational Intelligence in Data Mining 711: 605-16. https://doi.org/10.1007/978-981-10-8055-5_54.

Liliia Z (2020) Analysis of statistics of pancreatic cancer in Ukraine for a period of 10 years. Journal of Clinical Oncology 38 (15_suppl): e16721. https://doi.org/10.1200/JCO.2020.38.15_suppl.e16721.

Martinez-Duncker I, Salinas-Marin S, Martinez-Duncker C (2011) Towards in vivo imaging of cancer sialylation. International Journal of Molecular Imaging 2011: 1- 10. https://doi.org/10.1155/2011/283497.

Podgorsak EB (ed) (2005) Radiation Oncology Physics: A Handbook for Teachers. And Students. International Atomic Energy Agency, Vienna.

Podgoršak EB (2006) Radiation Physics for Medical Physicists. Part of the book series: Biological and Medical Physics, Biomedical Engineering (BIOMEDICAL). Springer. ISBN: 978-3-540-29471- 9.

Podgorsak EB (2009) Biological and Medical Physics, Biomedical Engineering. In Radiation Physics for Medical Physicists. 2nd enlarged edn. Springer Heidelberg Dordrecht London New York. ISBN 978-3-642-00874-0. e-ISBN 978-3-642-008745-7. https://doi.org/10.1007/978-3-642-008745-7.

Podgorsak EB (2010) Radiation Physics for Medical Physicists (Biological and Medical Physics, Biomedical Engineering). Springer-Verlag Berlin Heidelberg. ISSN 1618-7210. ISBN 978-3-642- 00874-0. e-ISBN 978-3-642-008745-7. Doi 10.1007/978-3-642-008745-7.

Podgorsak EB (2016) Radiation Physics for Medical Physicists. Springer. ISBN: 978-3-319-25382-4.

Schreiber GJ (2022) Meyers AD (ed) General Principles of Radiation Therapy. Medscape.

Shin-Ichi M, Shinji K, Ryo H, Toshihiko K, Minoru S, Natsuko K, Koji O (2016) Boron neutron capture therapy for malignant brain tumors. Neurol Med Chir (Tokyo) 56 (7): 361-371. Smith Y (2021) Radiation Therapy Mechanism. News Medical Life Sciences.

The 2007 Recommendations of the International Commission on Radiological Protection. Annals of the ICRP. ICRP publication 103: 37 (2–4). ISBN 978-0-7020-3048-2.

Tofilon PJ, Camphausen K (eds) (2017) Increasing the Therapeutic Ratio of Radiotherapy. Cancer Drug Discovery and Development. Humana Press. Springer. ISBN: 978-3-319-40854-5. https://doi.org/10.1007/978-3-319-40854-5.

Wyleżoł M (2017) Methods of Skull Implants Modeling with Use of CAx and Haptic Systems: Medical Imaging: Concepts, Methodologies, Tools, and Applications 22. IGI Global. https://doi.org/10.4018/978-1-5225-0571-6.ch02.

Zhu A, Lee D, Shim H (2011) Metabolic PET imaging in cancer detection and therapy response. Semin Oncol 38(1): 55–69. https://doi.org/10.1053/j.seminoncol.2010.11.012.

Zhu A, Marcus DM, Shu H-KG, Shim H (2012) Application of metabolic PET imaging in radiation oncology. Radiation Research 177(4): 436-448. https://doi.org/10.1667/RR2702.1

Metabolic and Biological Effects of Deposited Radionuclides

Key Learning Objectives

- Metabolic effects of radionuclides
- Pharmacological studies
- Toxicological studies
- Analytical applications
- Biological effects of radiation

Overview

As already discussed in the previous chapters, radioisotopes are used in several medical applications, i.e., in the diagnosis and treatment of diseases. Studies of the fundamental processes in the body may not have immediate diagnostic or therapeutic applications. However, these studies can indirectly result in practical applications. The radioisotopes are routinely used now in a variety of biochemical processes that can be traced and quantified in the modern laboratory, i.e., they are useful in following the trajectories of biochemical reactions and ascertaining the distribution of a substance in an organism. In biomedical research, radioisotopes have also proven to be crucial tools. The field of drug metabolism and pharmacokinetics has been greatly impacted by recent developments and the usage of radioisotopes in designing and carrying out ADME investigations. Apart from this, the radiations also have an effect at the cellular level and thereby have varied types of consequences on the body.

A molecule that is identical in chemistry and function to the natural molecule of interest and to which stable or radioactive isotopes are attached is called a *radiotracer*. There is a long list of radioisotopes, but the isotopes useful in research are

A. Pathak, *Tools and Techniques in Radiation Biophysics*,
https://doi.org/10.1007/978-981-99-6086-6_11

very few. An isotope must meet certain criteria so that it can be used in research. These criteria are that it must already be a component of the experimental system so that it gets incorporated into those molecules and then used in tracing the path of biochemical reactions, as metabolic tracers to trace the diffusion of a molecule within a cell or tissue, and as a flow tracer to monitor fluid flow. For metabolic research, isotopes of carbon, hydrogen, oxygen, and phosphorus are popularly used. In some cases, an element can also be substituted for another; for instance, isotopes of sulphur can take the role of oxygen.

1 Metabolic Effects of Radionuclides

Exposure to radiation may alter the normal metabolic processes of the body, thereby inducing cellular responses that may alter the functioning of the body. The following are the various effects of radiation on metabolism:

1.1 Metabolic Pathways

Radioisotopes are routinely used to track down and measure a number of biological processes. These isotopes have been extensively employed to study every aspect of metabolism. For example, amino acid metabolism, biosynthesis of protein, degradation of carbohydrate, TCA cycle, nucleic acid synthesis, heme biosynthesis, fatty acid degradation and biosynthesis, cholesterol and steroid metabolism, photosynthesis, etc. The use of isotopes has made it possible to distinguish and identify metabolites from significant metabolic pathways. These are frequently used to trace metabolic pathways and for measuring the flow of metabolites through them. This involves the addition of a radioactive substrate, then taking samples of the substance at different time intervals and separating the products formed by HPLC, gas chromatography, or thin layer chromatography. The detectors measuring radioactivity can be attached to chromatography columns to detect and measure radioactivity coming from the column during separation. Radioactivity can also be detected by thin layer chromatography with either a GM scanner or by autoradiography (already discussed). These techniques allow for finding out the fate of each individual atom in a biological compound. These studies can be performed by using radioactive carbon, hydrogen, phosphorus, iodine, and others according to the need and constraints of the experimental procedure. Moreover, minor metabolic pathways that are converted into major pathways in disease have also been determined with the help of isotopes. Most of the metabolic pathways known today largely owe their elucidation to these techniques.

For example, radioisotopes are used to study glucose catabolism. There are a number of ways in which glucose gets oxidised: glycolysis, the TCA cycle, and the pentose phosphate pathway. Most of the enzymes involved in all these pathways are present in the body. All these pathways lead to the total oxidation of glucose to CO_2, but the origin of the CO_2 in terms of the 6-carbon atoms of glucose varies (at least during the initial phase of respiration). Therefore, it is possible to find out the contribution of each carbon atom to glucose oxidation. For this, the CO_2 released during respiration is trapped using glucose labelled with 6-^{14}C and 1-^{14}C glucose.

Radioactive tracers have been used to trace and assay enzymic reaction and a large number of methods have been developed for these studies. One example of this is the pulse-chase technique, where a radioactive substance is introduced into the biological system as a pulse, and the radiotracer is chased through various metabolites or cellular compartments.

The ability of the radioisotopically labelled metabolites to function like their non-radioactive counterparts and, hence, to be traced has permitted the researchers to study virtually every aspect of cell metabolism in all organisms including bacteria, yeasts, insects, plants, animals, and humans. Such studies are useful in understanding similarities in the metabolic processes of organisms across the evolutionary tree; the identification of minute differences in the structure and function of cells between organisms and, consequently, the ability of the drugs to destroy the infecting organisms, including bacteria, fungi, or insects, without harming humans; and the explanation of the fundamental characteristics of genetic material. This is because of basic advancements in molecular biology and genetics over the past 50 years, which significantly relied on studies using lower organisms along with radioisotopically labelled materials. This development has important indications in the present scenario, as the genes controlling important body functions are being recognised and cloned. Gene therapy is just beginning to discover its way into several clinical applications, as research on both human and nonhuman models using radioisotopes has shown a positive effect on human health.

1.2 Metabolic Turnover Times

Metabolic turnover is the aggregation of all biochemical activities taking place in the living cell that help in converting substrates, mainly carbon, nitrogen, and energy sources, into secreted metabolites, free energy, and cell mass. It is also represented as the average time during which the entire reservoir of a particular metabolite will be replenished due to production and utilisation rates. Radioisotopes provide an appropriate method to discover the turnover times of particular compounds. For example, the radioactive isotopes ^{14}C and ^{3}H help in the measurement of the turnover of substrates used for the metabolism of carbohydrate and protein. The tracer can be administered intravenously, and its metabolites can be monitored by monitoring the levels of tracer in the blood, breath, and urine. For example, a ^{14}C-labelled metabolite helps in determining the oxidation rates by calculating the decreased rate of $^{14}CO_2$. Turnover of proteins can be studied using ^{15}N-labelled amino acids; these were fed, and they emerged in polypeptide chains in liver proteins at a reasonably high rate even when the total amount of protein remained the same. If instead of ^{15}N-labelled amino acids, ^{14}N-labelled amino acids were fed, the earlier labelled liver proteins would also rapidly lose their label without any change in the total quantity of liver protein. It can be concluded from the experiment that the proportionately high rate of formation and degradation of proteins in the liver balance each other. Thus, metabolic turnover occurs in liver proteins. Such experiments help in measuring the half-life and, hence, the turnover rates of proteins in different organs. Similarly, the metabolic turnover times of carbohydrates, lipids, and other cellular constituents can be obtained. Another example is the quantification of brain blood turnover using radioisotopes. This serves as clinical evidence of brain circulation. This was carried out using gamma-labelled hippurate injected intravenously and measuring the time period of the tracer through the brain blood pool by externally monitoring the cranial passage of a bolus of radiotracer. Radioisotopes are also useful to study the effect of weightlessness during space travel on the human body. Radioisotopes have made it possible to observe the consequences of space travel on blood plasma volume, total body water, extracellular fluid, red cell mass, red cell half-life, and bone and muscle turnover rates more accurately. Metabolic turnover rate is also useful in understanding the management of metabolic pathways in plants. Photosynthesis in the leaves of vascular plants involves sugar phosphate metabolism and the Calvin cycle. So a ^{13}C-labelled system was used to determine the metabolic turnover rate in such plants.

1.3 Mineral Metabolism

Minerals are crucial for the human body. They form the components of several proteins and enzymes. They are crucial for the effective operation of cells, tissues, and organs and play significant roles in metabolism and body functions. Radioisotopic tracers are more sensitive, accurate, and fast for estimating the concentration of nutrients and mineral metabolism, that includes the site of minimal absorption, their distribution and excretion in the digestive tract, their kinetics in the blood, their distribution in different tissues and organs, and their excretion pattern from the body. Extensive data on the bioavailability of minerals in humans and animals has been obtained by using isotopic tracers. For example, iron metabolism in the body can be studied using radiotracers. Iron forms an important constituent of haemoglobin that transports oxygen from the lungs to all body cells. Studies with radioactive iron confirmed the pathway that iron follows from being ingested through food to being used in haemoglobin before being excreted from the body. These investigations had real-world implications for nutrition, blood disorders, and the significance of iron metabolism during pregnancy. Therefore, radiotracers have made a major impact on the understanding of mineral metabolism in animals and have been and continue to be employed extensively for medical diagnostic purposes in human and fairly extensively in the study of mineral metabolism in adults. In infants, radiotracers have been used for studies like the measurement of RBC survival using ^{51}Cr-labelled RBCs or for issues related to blood loss using either ^{51}Cr- or ^{59}Fe-labelled RBCs. ^{50}Cr-labelled RBCs have been used to study RBC survival time. Metabolic studies of calcium and phosphorus have been extensively done with this technique using ^{45}Ca and ^{32}P, respectively. From the tracer experiments with radioactive phosphorus, it has been concluded that vitamin D acts to help the formation of inorganic phosphorus in the bone. Using radioactive calcium and strontium, it has been concluded that vitamin D has a direct impact on the process of bone mineralisation as well as encouraging calcium absorption from the digestive system, all of which contribute to its ability to promote healing. It is also possible to measure the absorption of calcium from the gut and the rate of endogenous secretion with the help of tracer techniques.

1.4 Studies on Assimilation, Accumulation, and Translocation

Radioisotopes have been frequently used to study the mechanisms and rates of absorption, accumulation, and translocation, i.e., change of site of the absorbed substance, in the case of both organic and inorganic materials by both plants and animals. Absorption studies are done to find out whether some of the essential nutrients are absorbed from the gut, or not and if they are absorbed, how much is being absorbed. These types of studies have a role in detecting several diseases like pernicious anaemia, where vitamin B12 is not absorbed. This will also help in finding the route, translocation, and the site of accumulation of the molecules of biological interest. This technique has been extensively used to study calcium metabolism. The calcium and phosphorus exchange reactions that occur in the bone have been obtained from studies using ^{45}Ca and ^{32}P. Isotope-labelled fats, proteins, and vitamins are also being used for the study of intestinal absorption. Radioactive atoms of carbon, phosphorus, and sulphur can be used to study several features of the metabolism of growth regulators in plants. These have also been used to detect even tiny amounts of insecticide in the plants. It has also been feasible to track any chemical via absorption in the leaf or root and translocation through the plant tissues. Most autoradiographic methods have been used for such studies.

1.5 Pharmacological Studies

The use of isotopically labelled pharmaceuticals has significantly expanded the range of pharmacological research. Radiotracers also find use in the development and formation of new drugs. Radioisotopes are labelled with the drugs, and they are mainly used in the investigation of their pharmacological kinetics, i.e., determining the location of drug accumulation, the rate of accumulation and metabolism, and the formation of metabolic products, etc. Pharmacological studies can be accomplished by incorporating a radioisotope into the drug. Such studies help in understanding the features of new drugs with regard to their clinical applications, i.e., how a drug traverses through the blood to different tissues and undergoes metabolic transformation in structure. Only after a detailed study can it be sent to the FDA for approval and then be used by the general public.

Radioisotopes are also used in the research and development of pharmaceuticals. These are often used in clinical studies for the development of compounds for various therapeutic areas, including diseases related to the central nervous system (such as dementia, Alzheimer's disease, and Parkinson's disease), oncology, and metabolic diseases (like diabetes mellitus).

A variety of radioisotopes are used in the research and development of drugs, such as ^{3}H, ^{14}C, ^{32}P, ^{35}S, and ^{131}I, because these isotopes can be easily incorporated into many drugs. These radiolabelled drugs can be comfortably detected and measured, and the disposition of drugs can be easily assessed using liquid scintillation counters. Tagging with stable radioactive isotopes is one of the most suitable methods for the estimation of pharmacokinetics and structure elucidation at the same time in drug metabolism studies. The radiopharmaceutical studies depend on the type of radioisotope, the placement of the radiolabel in the drug molecule, the purity of the radiopharmaceutical, and its specific activity because these parameters affect the metabolism, and stability of the radiopharmaceutical, as well as the detection and recovery of radioactivity.

Radioactively labelled compounds are also used in certain non-clinical studies to evaluate drug absorption, distribution, metabolism in the body, and excretion. They provide information on

(a) Drug uptake in cell lines,
(b) P-glycoprotein (P-group) transport or inhibition,
(c) Metabolic profiling,
(d) Drug transport (i.e., uptake efflux), and
(e) Drug attachment to receptors.

1.6 Toxicological Studies

Toxicology is the study of xenobiotics, i.e., foreign chemicals entering the body. The fate of xenobiotics *in vitro* and *in vivo* was developed initially by using radioisotopes. The development of techniques for the detection and quantification of energy released by radioisotopes led to the development of synthetic xenobiotics containing radioactive isotopes so that they could be tracked in organisms.

The most prevalent application of radiolabelled xenobiotics in toxicology is the study of the fate of a foreign chemical in animal models. These studies are called absorption, distribution, metabolism, and excretion (ADME). A successful ADME study needs:

- A radiolabel xenobiotic with the appropriate type and position of the label,
- An analytical technique for segregating the parent compound from its postulated metabolites, and
- Methods for quantifying the amount of radioactivity.

The simplicity of detection and the attainment of high sensitivity are the main benefits of using radioisotopes to study the disposal of new drug candidates.

The radioisotope of choice for xenobiotic ADME studies is ^{14}C because it is the radioactive form of the element that serves as the basis for the majority of xenobiotics. In addition, ^{14}C has a long half-life of more than 5000 years and is a week beta emitter, so it poses fewer health risks to the workers. To prevent the label from being lost during metabolism, the label's position must be meticulously chosen. ^{3}H, a radioactive form of hydrogen, is also used as a tracer because it has high specific activity. This allows researchers to detect smaller amounts of xenobiotics. There are several other radioisotopes that have also developed applications in studying the metabolic disposal of compounds, like ^{11}C, ^{14}C, ^{36}Cl, ^{125}I, ^{131}I, ^{32}P, ^{33}P, ^{35}S, ^{87}Ru.

Sometimes a xenobiotic itself is not as toxic as its subsequent metabolites may prove. Then radiolabelled xenobiotics may be taken to study the metabolic pathway and ascertain the significant culprit causing pronounced toxicity.

1.7 Analytical Applications

1.7.1 Research on Enzyme and Ligand Binding

Radioisotopes have also been used to study the mechanisms of enzyme action and ligand binding, i.e., the binding of ligand to membrane receptors. Radiotracer methods can be used to study effectively any enzyme reaction on the condition that a radioactive form of the substrate is available. Radiotracer-based enzyme assays are more expensive, but they offer a very high level of sensitivity.

A radioligand is a substance that is labelled with radioactivity and can bind to a receptor, transporter, enzyme, or other site of interest. It should ideally have high affinity, high receptor specificity, low non-specific binding, and high specific activity to detect low receptor densities. Measuring the rate and degree of binding reveals the number of binding sites as well as their affinity and other pharmacological properties. Radioligand binding is an approach that makes use of a radioactivity-labelled compound, which binds at the target binding site. Nowadays, these binding assays are used for studying membrane-bound molecules, such as G-protein-coupled receptors. They are also used to determine the type of receptors and their anatomical distribution and density. Radioligand binding studies are also a crucial method for quantifying and quantitatively characterising receptors and have thus become a standard method in a variety of scientific disciplines. These assays are suitable for purified proteins, tissue homogenates, cell lysates, and whole cells.

Radioligand binding can also be used to:

- Identify receptors in their innate environment and those transfected into cell lines,
- Analyse receptor dynamics and localisation,
- Identify unique chemical structures that interact with receptors,
- Describe ligand activity and discrimination in healthy and pathological tissues, and.
- Define receptor functions at the molecular level.

Three main experimental protocols are involved in these radioligand binding studies. This way, the binding kinetics of the competitor (or test compound) are determined without the need to label each individual compound.

1. *Kinetic experiments*: Here the time path of radioligand binding and dissociation of radioligands from the protein is measured. Kinetic investigations (i.e., association and dissociation) are necessary to determine the optimal assay conditions. Theoretically, they also enable the most accurate determination of the radioligand's affinity for its corresponding receptor. However, kinetic studies are inconvenient to perform and therefore hardly used in routine studies.
2. *Competitive binding experiments*: They are used to assess the affinity of the receptor for the competitor by measuring the binding of a single concentration of radioligand at different concentrations of an unlabelled competitor.
3. *Saturation binding experiments*: Here, the extent of binding is assessed for the presence of several concentrations of the radioligand that ultimately helps in determining the number of binding sites and ligand affinity. Fractions of partially purified plasma from tissue homogenates, cells transfected with cloned receptors, and cells that are either in culture or extracted prior to analysis—all these can be analysed by saturation binding experiments.

There are two kinds of approaches adopted for such experiments.

- Increasing the quantity of radioligand added while maintaining the specific activity and concentration of radioligand constant, or
- Decreasing the specific activity of the radioligand as a result of the addition of an unlabelled ligand.

1.7.2 Radioanalytical Methods

The principal radioanalytical methods often used in biology and medicine are grouped under three main headings:

1. Isotope dilution analysis.
2. Activation analysis.
3. Labelled reagent method (including derivative analysis and immune radiometric assay).

Isotope Dilution Analysis

This is a quantitative analytical technique for determining the mass and quantity of chemical substances. The method consists of adding a known amount of isotopically enriched substance (i.e., an isotopic standard) to the sample being analysed. The mixing of these two dilutes the isotopic supplement of the standard, and this forms the foundation for the isotope dilution method.

There are three main types of isotope dilution methods: (a) direct, (b) inverse, and (c) double isotope dilution. The fundamental principles on which all these methods are based differ, but they differ both in technology and procedure and are applied under different circumstances.

This method can provide high sensitivities and therefore can be used to assay those compounds that are present in low amounts in living organisms and in mixtures of similar compounds accurately and conveniently. The benefit of the isotope dilution method is that after mixing the spike and sample, the dose in the substance has no longer an impact on the analysis's outcome.

There are a number of applications of isotope dilution analysis in biochemistry and medicine. This method is mainly employed in analytical chemistry for accurate quantitative analysis. This is important in geophysical investigations, for example, in determining the presence and distribution of trace metals in geological samples and meteorites. This is also used in studies including measurement of erythrocyte volume with ^{51}Cr-labelled erythrocytes; plasma volume measurement using ^{131}I-labelled serum albumin; evaluation of body water using ^{131}I-labelled iodoantipyrene or tritiated water; estimation of extracellular space with ^{82}Br; calculation of total body pools of cholesterol, urea, and uric acid using ^{14}C; and sodium and potassium pool assessment with ^{24}Na and ^{42}K, respectively.

Isotope dilution assays of a wide variety of substances, such as vitamins, antibiotics, and insecticides, have also been developed.

Activation Analysis

This method of analysis is relatively new and is being increasingly employed and developed. This is an analytical tool for calculating the concentration of trace metals in a sample. It is a nuclear method of qualitative and quantitative determination of metals in a wide variety of substances. The sample is made radioactive by some artificial means, i.e., by bombardment with appropriate nuclear particles, such as neutrons, charged particles, or photons (gamma rays). The energy linked with the radiation is specific to the radioactive isotope used, and therefore, the artificially produced radioisotopes are recognised and the elemental quantitation done depending on the gamma rays they emit. Activation analysis is established for nuclear reactions that produce radionuclides peculiar to a given element. After irradiation, these radionuclides are quantified. Both qualitative analysis (i.e., identification of the radiation emitted on the basis of its energy and half-life) as well as quantitative analysis (i.e., measuring radioactivity) are possible. The significant analytical feature of activation analysis is its innate sensitivity, that is suitable for elemental analysis at values as low as 10^{-12} g for certain species. Activation analysis is applicable to most types of materials, including food products, urine, faces, air pollution particulates, river water, marine samples, vegetation, soils, sediments, ores, plastics, petroleum products, pharmaceuticals, coals, metals, alloys, ceramics, and glasses.

Different types of activation analysis methods exist that are distinct from each other on the basis of the following four parameters:

- The type of radiation or particle bombardment used for sample excitation,
- The type of radiation involved in the final emission process,
- Sample preparation by isolation of the species of interest, and
- Direct sample monitoring without separation using gamma ray spectrometer.

The most frequently used particle is the neutron and the method is then known as neutron activation analysis. Elemental investigation by the neutron activation method is done using neutrons produced by a reactor that form radioactive isotopes from the stable ones in a sample material. Both beta particles and gamma rays are released from the radionuclides during their decay. These gamma rays have distinct energies for a specific radionuclide, and the rate of emission of this radiation can be determined using gamma spectroscopy or high- resolution semiconductors. This rate of emission of gamma rays from an element in a sample is directly related to the concentration of that element. This method of analysis is capable of detecting the nature and quantities of various elements present in the sample at extremely low concentrations. This is a multi-element method, i.e., more than 70 elements can be detected in the sample simultaneously. This is important in examining lunar samples, artefacts, and criminal investigations. It is a non-destructive method, i.e., analysis can be carried out without loss of the sample due to the absence of defects associated with yield determinations. Analysis can be done on a sample as it is received. In most cases, no chemical treatment is needed to prepare the sample for analysis. Therefore, contamination caused by excessive sample handling and reagent addition is eliminated. Neutron activation analysis is also useful in verifying trace element concentrations in complex matrices like human and animal tissue, coal, coal fly ash, petroleum, river sediments, waste water, geological deposits, paper, air particles, etc.

Therefore, neutron activation analysis is a useful, sensitive analytical technique for undertaking qualitative and quantitative multi-element analyses of major, minor, and trace components in a variety of terrestrial samples and extraterrestrial substances. The detection limit ranges from 0.03 ng to 4 ng. It is a highly precise method that gives accurate results such that an overall error of only 2–5% is achieved. Samples as small as a few micrograms can be analysed by neutron activation analysis. This is generally recognised as the "referee method" of choice.

The charged particle activation analysis method is another method that is based on reactions induced by charged particles. High-energy deutrons, protons, or helium particles are bombarded with the sample to generate radioactive nuclides inside the analysed material. The incident charged particles interact with an element to produce radioactivity whose intensity is proportional to the concentration of the element and whose concentration is determined by direct measurement of the induced radioactivity. This method is especially useful for the most sensitive and precise analysis of lighter elements like oxygen, boron, and nitrogen. This technique is extremely sensitive; however, the beam has a very small depth of penetration and therefore can be used to analyse only the surface of the sample. This property makes it different from other activation analysis techniques.

Photon activation analysis makes use of photon-induced reactions, i.e., samples are irradiated with high-energy photons to produce radionuclides that are then measured by gamma ray spectroscopy. Like neutrons, these energetic photons are able to penetrate thick, solid samples. This allows bulk investigation without chemical pretreatment. However, a large sample mass is required for analysis.

Labelled Reagent Method

^{125}I-labelled antibodies are used that bind to specific proteins, peptide hormones (for example, thyroid and steroid hormones), viral antigens, etc. These methods are collectively termed immune radiometric assays.

Radioimmunoassay (RIA)

> **Key Concept**
> Yalow and Person developed the first radioimmunoassay assay during their work on insulin and were awarded the Nobel Prize in medicine in 1977.

RIA is based on the competition between labelled and unlabelled antigens for specific antibody sites, forming antigen-antibody complexes. This is an extremely sensitive *in vitro* technique that is used to determine the concentration of antigens (for example, levels of hormones in the blood) by using antibodies against these antigens. In RIA, a known quantity of antigen is labelled with a radioactive substance and combined with a known quantity of antigen-specific antibody. The non-radioactive antigens take the place of the radioactive antigens when the sample containing the target antigen is added. After washing away non-fixed antigens, it is feasible to measure the ratio of radioactivity in the supernatant to that of the standards. Therefore, it is feasible to measure the amount of antigen in the sample. RIA has transformed scientists' ability to measure and quantify small amounts of tissue components, like enzymes, hormones, or serum proteins, by determining the component's ability to attach to an antibody or other proteins in competition with a known amount of the radioactively tagged component (Fig. 1). This technique has enabled the diagnosis of numerous human diseases without exposing patients to radioactivity. It is highly sensitive and specific and is used to measure concentrations of substances, usually antigen concentrations (for example, hormone levels in

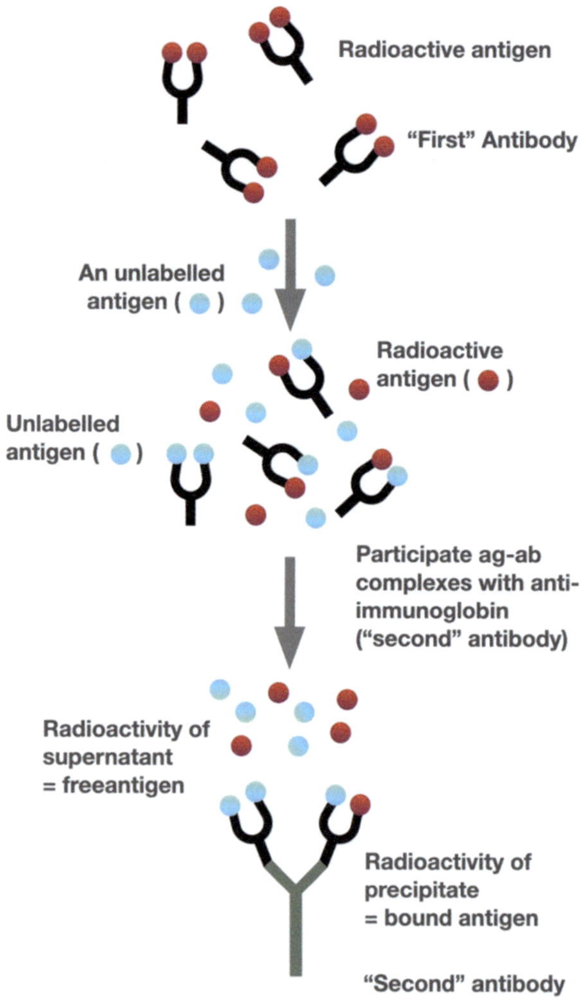

Fig. 1 Principle of radioimmunoassay

the blood) by using antibodies. This method has transformed research and clinical practice in many areas, for example, blood banking, diagnosis of allergies, endocrinology, etc.

This technique offers a number of advantages:

- *Sensitivity*, i.e., it is able to detect very little amounts of substances because of its ability to detect very low amounts of radioactive tracers. By using high-affinity antibodies, it is possible to detect even a few picograms (10^{-12} g) of antigen in a test tube.
- *Specificity*, i.e., the ability to measure only the substance of interest in a complex sample. The specificity of the assay in proportional to the specificity of the antiserum.
- *Accuracy and reproducibility*, i.e., to determine the effective quantity of a substance precisely and generate highly reproducible results of concentrations.

The main drawbacks of RIA are the cost and hazards of producing and handling the radioactive antigen.

Despite these drawbacks, RIA has become a major tool in a clinical laboratory, where it is used in the assay of

- Plasma levels of

  Most of the hormones, and
  Digitoxin or digoxin in patients on these drugs

- Certain abused drugs,
- For the presence of hepatitis B surface antigen (HBsAg) in donated blood, and
- Anti-DNA antibodies in systemic lupus erythematosis.

Immunoradiometric Assay

Immunoradiometric assay (IRMA) is an assay that uses radiolabelled antibodies in place of antigens. It is a non-competitive radioimmunoassay (RIA). It differs from the conventional radioimmunoassay (RIA) in that here the compound to be evaluated combines at once with radiolabelled antibodies rather than replacing another antigen. In IRMA, the antibodies are tagged with radioisotopes that bind the antigens in the specimen. When a positive sample is put in the tubes, radioactively labelled (for example, labelled with ^{125}I or ^{131}I) antibodies bind to the free epitopes on antigens and form an antigen-antibody complex. A second reaction with a solid phase antigen removes unbound, labelled antibodies. The quantity of radioactivity remaining in the solution is directly proportional to the antigen concentration.

High sensitivity is an advantage of IRMA, but radioactive contamination is its main flaw. After monoclonal antibodies became available, a new form of IRMA known as the two-site or sandwich IRMA was developed. In this method, two antibodies specific for the same antigen but binding to distinct epitopes are used. One of the antibodies is deposited onto the solid phase and used as an immune extractor for the antigen. A second antibody labelled with ^{125}I is added to the solid phase. This labelled antibody binds to the antigen that is already bound to the first antibody. At the conclusion of the reaction, the unreacted antibody is aspirated out, and the solid phase is rinsed with a wash solution. The radioactivity in the solid phase is proportional to the amount of analyte present. The two-site IRMA is more specific than the RIA because it employs two antibodies. CLIA (chemiluminescence immune assay) is a mature and advanced technology that has developed rapidly over the past 10 years and has a broad range of applications using fluorescent labelling. It is more sensitive and precise. It is anticipated that CLIA will replace the majority of IRMA and ELISA methods in clinical settings.

Radiometric Dating

Key Concept
Bertram Boltwood first published a paper on the use of radiometric dating in 1907.

Radiometric dating is a type of absolute dating that gives an exact age for how old a material or object is. It is a method for dating materials such as minerals and carbon in which trace radioactive impurities were selectively incorporated during their formation.

This is done by measuring the amount of radioisotope left (or by examining the quantity of the decay product) from the understanding of the half-life. For example, if the radioisotope comprises 1% of the element and the sample actually has 0.25%, then two half-lives can be assumed to have passed since deposition. If the half-life of the isotope is one million years, then the sample can be dated as being two million years old.

There are four main radiometric dating methods:

- *Uranium-lead dating*- used to date uranium-bearing rocks and minerals. Two different unstable isotopes of uranium decay into stable isotopes of lead through a decay chain of alpha and beta decay.
- *Potassium-argon dating*- used to date potassium-bearing rocks and minerals. Unstable potassium decays into a stable isotope of argon through beta-decay and the electron capture process. This is the most extensively applied method of radiometric dating, as potassium is a component of many usual minerals and, therefore, can be used to ascertain the ages of igneous and metamorphic rocks.
- *Rubidium-strontium decay*- used for dating rubidium-bearing rocks and minerals. Unstable rubidium decays into a stable isotope of strontium through beta decay.

- *Radiocarbon dating*- used to date objects that were once living and have elemental and isotopic carbon within them. For example, wood, cloth, fossils, or human remains. ^{14}C decays into ^{14}N through beta decay.

Radiometric dating measures the amount of radioactive material left in an object and the amount of decay product in the object. The ratios are compared, and calculations are done using the known half-life value to find the age of the object.

The advantages of radiometric dating include:

- the capacity to date an object without damaging it,
- having a wide range of techniques to choose from, and,
- the ability to determine a relatively precise age of objects that are hundreds, thousands, millions, or billions of years old.

Geologists can document the history of the earth and its events, such as the dinosaur era, using radiometric dating on a geological time scale. However, it can only be done on certain materials and is not helpful for measuring the age of sedimentary rocks. All types of radioactive dating except ^{14}C are based on atoms found in igneous rocks. Fossils are never found in igneous rocks. So radiometric dating cannot be used directly to date fossils. ^{14}C can be an option. However, ^{14}C has a very short half-life. So it can't be used to date fossils.

Another limitation is that the material being dated must contain detectable levels of both the parent and the daughter isotopes.

Check Points

- What are radiotracers?
- How are radioisotopes used to study metabolic pathways?
- What is metabolic turnover time?
- Explain the effect of radionuclides on mineral metabolism.
- What are radioligands? Explain their importance.
- Describe isotope dilution analysis.
- Explain the different types of activation analysis methods.
- Write a note on the radioimmunoassay.
- What is an immunoradiometric assay?
- What is radiometric dating? Explain the different methods of radiometric dating.

Thus, it can be concluded that ionising radiation exposure disturb normal metabolic processes in cells and organs by triggering complex biological responses. These effects are dependent on the dose and the rate of exposure. Radioisotopes are being used these days to study changes in metabolism, toxicological studies, and pharmacological studies.

2 Biological Effects of Radiation

Biological effects are caused in humans and other living organisms by their radiation exposure. These effects on living cells may result in three consequences:

- Injured or damaged cells restore themselves, resulting in no residual damage.
- Cells perish and are replaced by normal biological processes.
- Cells mistakenly repair themselves, resulting in a biophysical change.

2.1 *Distribution Studies*

The study primarily consists of

(a) administering a radioisotope to a system or organism, and,
(b) localisation of the isotope through subsequent detection of radioactivity in various parts of the system or organism.

Such studies can be quantitatively conducted, and the distribution's time course can be measured. The distribution may be measured at several levels of the organisation. The detection level may be macroscopic (for example, a whole organ) or microscopic (for example, subcellular particles like the mitochondria, ribosomes, or the nucleus).

Distribution studies can be used in the following biological and medical research:

1. Iodine is an important element in the thyroid physiology. ^{131}I administered as NaI has given information about thyroid physiology. The absorption of ^{131}I by the thyroid gland and the appearance of the isotope in saliva and urine have been used to measure thyroid functions.
2. Malignant tissues have an accelerated metabolism and take up more phosphate than normal tissues. Consequently, if ^{32}P is injected into a patient, after a certain time, more ^{32}P will accumulate in the tumour, if it is present. This can be found by placing a photosensitive emulsion in contact with the tissues. The malignant tissue will show up as a shot blocker, unlike the normal tissues.

2.2 *Effects of Radiation on Humans*

High radiation doses are known to kill cells, but low doses tend to destroy or alter the genetic code of irradiated cells. High doses affect so many cells that tissues and organs are immediately damaged. Low doses (100 mSv) spread out over long periods of time (years) and don't cause immediate trouble for any body organ. The effects of low doses of radiation take place after several years. Therefore, the biological effects of radiation can be organised into somatic and hereditary effects. Somatic effects are radiation-induced tissue or health effects that are not heritable,

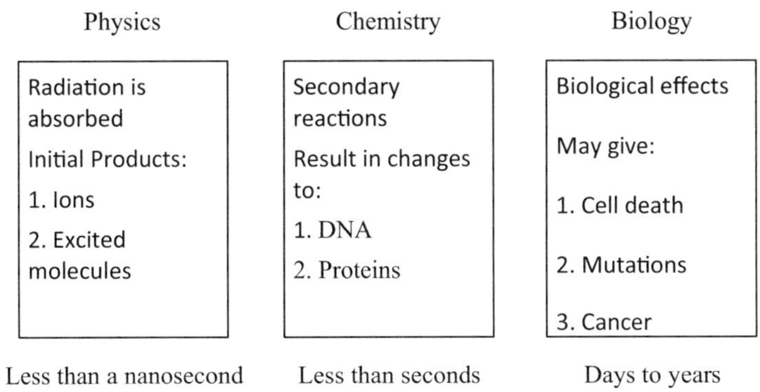

Fig. 2 Some of the products and secondary reactions that take place when a biological system is irradiated

while genetic effects are caused by radiation- induced mutations of any germ cell that lead to lasting changes in genetic material that are therefore inherited by offspring of the exposed individual. Depending on the level of radiation exposure, excessive radiation exposure may harm living tissues and organs. The extent of potential injury depends on a number of variables, including:

- the type of radiation,
- the sensitivity of the affected tissues and organs,
- the method and duration of exposure,
- the radioactive isotopes involved, and
- characteristics of the exposed individual (i.e., age, gender, and underlying condition).

These are some fundamental mechanisms for the action of radiation on biological molecules ranging from water to DNA. The initial process is described in Fig. 2 below.

Key Concept
DNA is the primary target for biological damage.

Exposure to radiation leads to the formation of a number of "primary products". These products (for example, ions, excited molecules, or free radicals) are highly reactive, with lives of the order of a fraction of a second in a cell. This reaction with molecules in the cell results in secondary processes that ultimately yield observable results like cancer, genetic change, or cell death.

Radiation can have serious consequences when it comes into contact with human cells. When radiation traverses matter, it transfers some energy to the absorbing medium through either ionisation or excitation of the atoms. It is the ionisation of atoms in tissues, accompanied by chemical changes, that causes the harmful biological effects. Thus, all biological effects occur because of radiation interactions with the atoms in the cells. Radiation affects the cells by two mechanisms, i.e., direct and indirect effects.

Direct effect: If radiation interacts directly with the DNA molecule or some other cellular component important for the survival of the cell, it is called a direct effect. This effect is observed in the same molecule where the primary absorption occurs, i.e., with which the radiation interacts. This type of interaction may affect a cell's ability to reproduce and, therefore, survive.

Indirect effect: Here radiation is absorbed by one molecule (mainly water), resulting in the formation of reactive products (free radicals) that ultimately react with other molecules in the system. Radiation causes a highly coordinated sequence of events that are amplified by endogenous signalling and cause oxidative damage to DNA, lipids, proteins, and numerous metabolites. When ionising radiation traverses cellular tissue, it gives rise to charged water molecules. This occurs due to the ability of radiation to induce the ionisation of atoms. Radiation-induced ionisations can act directly on cellular component molecules or indirectly on water molecules, resulting in water-derived radicals.

$$H_2O + \text{Ionising energy} \rightarrow H_2O^* \tag{1}$$

H_2O^* represents an excited molecule.

The excited molecule is extremely unstable and rapidly loses its excess energy, resulting in the formation of highly reactive and chemically reactive H° and OH° (free radicals). These radicals strike nearby molecules immediately, resulting in the oxidation of the affected molecules, that can modify important molecules like deoxyribonucleic acid (DNA) in cells. These effects can lead to biological effects, including abnormal cell development and cell death (Fig. 3). The direct effect takes place in a dry system (without water), while in an aqueous system, the indirect effect predominates.

Fig. 3 Flow chart showing the mechanism of biological effects by charged particles

Charged particles

↓

Electrical interactions

↓

Ionisation

↓

Chemical changes

↓

Biological effects

If there are single-strand breaks in the DNA, they are repaired with no long-term effects, but if there are double- strand breaks, they are not easily repaired, and a long-term change will occur. However, the frequency of double strand breaks is relatively rare (1 double strand break in every 25 single strand breaks).

Let us discuss the effect of radiation on various molecules.

(a) **DNA**: Each cell contains chromosomes inside the nucleus that control the functions of every cell in the body. Chromosomes consist of two large DNA strands that make up the genetic code. When cells are exposed to ionising radiation, radiochemical injury can occur either directly or indirectly. Direct action takes place when alpha particles produce ions because of DNA breaks in a single strand or both strands. Single stranded breaks can be restored normally; however, repair of double-stranded breaks is much more difficult and may result in mutations, chromosomal aberrations, or cell death. Therefore, these are the most harmful lesions produced on chromosomes by ionising radiation.

   These repaired double-stranded breaks are cytotoxic and can result in the degradation of DNA fragments that can cause the joining of non-homologous chromosomes during repair, causing loss of amplification of chromosomal material. These events can lead to the formation of tumours (tumorigenesis). For example, if the region is deleted, it activates a tumour suppressor gene, or if the region is amplified, there is the formation of an oncogenic protein.

   Ionising radiation can also impair cells indirectly by forming free radicals that cause the formation of compounds such as H_2O_2, which initiate deleterious chemical reactions in the cells. Due to these, the cells may undergo several structural changes that can lead to altered cellular function and even cell death. In direct interactions, radiation causes dense ionisations along their tract. Therefore, the direct interaction between these energetic ions and DNA leads to its damage. This has a high level of biological consequence because the integrity of DNA sequence is important for biological function.

(b) **Amino acids:** Amino acids are the building blocks of proteins. Radiations induced the formation of free radicals of amino acids, i.e., °CHR-COO^- (deamination radical) and °CR(NH^+)-COO^- (dehydrogenation radical). These can be detected by ESR. Second impact of radiation on amino acids in molecular fragmentation (10^3–10^4 bonds are broken with energies between 10 and 100 eV). The common neutral losses are NH_3, H_2O, and COOH, that can be detected by mass spectrometry techniques. Ion irradiation can also lead to molecular cross-linking.

(c) **Peptides:** Homopolymers of the amino acids—glycine, alanine, valine, phenylalanine, and tyrosine—are being used as models to evaluate the effect of molecular structure due to the degradation of polypeptides by ionising radiation. The main reactions that can occur are:

   - scission of the entire side chain, resulting in its hydrogen adduct,
   - fragmentation of the side chain,
   - scission and fragmentation of the main chain.
   - More than one of these reactions may occur during ion irradiation-induced peptide fragmentation.

(d) **Proteins:** Proteins are another important biomolecules. Irradiation of proteins gives rise to specific structural and chemical alterations, including fragmentation, decarboxylation, aggregation, breakage of disulphide bonds, formation of disulphide radicals, release of aromatic amino acids, loss of helicity, etc. When proteins are degraded, a variety of products and several amino acids are produced. Damage to proteins occurs in a discrete manner. The capability of free electrons to traverse along the peptide backbone contributes to a mechanism for specific structural degradation to occur. Structural damage occurs on the basis of the bond strength, i.e.,

- Disulphide bond breakage,
- Decarboxylation of aspartates and glutamates,
- Loss of OH groups from tyrosines, and.
- C-S bond cleavage in methionines.

Damage can also occur to metal centres within protein, causing reduction. Active sites in proteins are also sensitive to injury, for example, the FAD cofactor in DNA.

A highly ordered three-dimensional structure is broadly accepted as the prerequisite for protein function. Radiation-induced protein damage like fragmentation, dephosphorylation, chain cleavage, and defolding will automatically cause protein denaturation, loss of function, and even cell apoptosis. On the other hand, proteins are also involved, either directly or as DNA-binding motifs, in several cellular activities such as DNA packaging, regulation of gene expression, and enzyme catalysis. Therefore, the lesions on proteins and their components mentioned above can not only denature proteins but also induce DNA damage and be lethal to organisms.

Although recently much attention has been paid to radiation-induced reaction cascades causing DNA damage, the protein lesions caused by energetic ions remain unexplored at the molecular level.

In several cases, acute exposure to radiation can lead to both immediate (radiation sickness) and delayed effects (cancer or death). Different types of tissues respond differently to radiation exposure. For example, bone marrow, gonads, the lining of the stomach, and hair follicles (active cells) are more sensitive to radiation than the brain or the skin (dormant cells). This is because the dividing cells need the correct DNA information for the survival of their offspring. A direct interaction between radiation and an active cell may cause cell death or mutation, whereas a direct interaction with the DNA of a dormant cell has less effect. Also, stem cells (germ cells) are more radiosensitive than mature, differentiated cells. So the most sensitive are haematopoietic cells, white blood cells (lymphocytes), epithelial cells of the intestinal tract, skin, and muscle cells, and the least sensitive are nerve cells. Cytological and molecular damage by radiation may be exhibited in the form of cancer, gene mutations, skin reddening, etc. Another important hazard is the accumulation of radiation in body tissues. Radioactive materials cause the most harm when inhaled as aerosols or ingested through contaminated hands, food, drink, and cigarettes. The risk of exposure depends on the complications of the procedure, the activity, and the radioactivity of the nuclide. Radionuclides are classified into four groups depending on their relative radiotoxicity:

- *Very high radiotoxicity*: This group comprises of plutonium and other alpha emitters. These materials are not routinely used in biomedical research laboratories.
- *High radiotoxicity:* These radionuclides include ^{125}I, ^{131}I, ^{45}Ca, ^{60}Co, and ^{137}Cs; all of these are not widely used in molecular research laboratories.
- *Moderate radiotoxicity (1 mCi to 1 Ci or 37 MBq to 37 GBq)*: These radionuclides include ^{14}C, ^{32}P, and ^{35}S. These are used for molecular biology research.
- *Low radiotoxicity (10 mCi to 10 Ci or 370 MBq to 370 GBq)*: This group comprises ^{3}H, ^{99m}Tc, natural uranium, and natural thorium. Out of these, only ^{3}H and ^{99m}Tc are relevant in the molecular biology laboratory.

Thus, the risk of developing unfavourable health effects varies with the radiation dose. The biological consequence of a given amount of radiation depends on several factors, including the total duration during which different doses of radiation are collected, the irradiation dose intensity and the category of irradiation. The possibility of adverse effects increases with dose. If the radiation dose is less or is given for a longer period of time, the risk is lower because the injury to cells and molecules will be simultaneously repaired by the body. Therefore, it is important to take strict precautions to safeguard consumers from the severe effects of radiation exposure.

2.3 Mechanism

There are two main categories of biological effects of radiation: deterministic and stochastic. *Stochastic effects* are produced when the frequency in the exposed occupants is directly related to the dose, regardless of how small the dose is. These effects are thought to have no threshold. Effects of this type are not fixed and may occur, but their probability increases with dose. However, the severity of an effect does not depend on the doses, for example, cancers of different types. *Deterministic effects* are caused when the severity of the exposure depends on the dose. Deterministic effects are produced only if the dose or dose rate is greater than a certain threshold value. So these effects are considered to have a threshold. Deterministic effects are mainly produced due to cell killing. These effects start to show up right away at the time of exposure and get worse at higher doses and dose rates. At excessively high doses, radiation can damage the functioning of tissues as well as organs and cause acute side effects including nausea and vomiting, hair loss, acute radiation syndrome, skin burns and redness, and sterility.

When ionising radiation penetrates a biological system, it loses its energy during the bombardment of complex molecules in the tissues. These collisions usually result in the ejection of electrons and the rupture of chemical bonds. This process culminates in the formation of molecular fragments, which are either ions or radicals. These ions, or radicals, due to their inherent instability, recombine in ways different from those in which they were formed. In this way, new molecules that are foreign to the biological system are formed. Usually, electrons ejected by the radiation contain sufficient energy to generate several other ions through bombardment. Radiation can thus set up a chain of events leading to the production of large amounts of ions, radicals, and new compounds foreign to the cell. More dangerous is the consideration that ionising radiation can interact with water, yielding free radicals, which are much more deleterious because they cause oxidation- reduction reactions on complex molecules that are necessary for the structural and functional integrity of a cell.

Radiation exposure produces different types of effects on different organs of the body. There are two main types of radiation health effects: (1) acute disorders and (2) late-onset disorders. Acute disorder is a deterministic effect, as the symptoms appear only with exposure above a threshold (already discussed above). The clinical symptoms of this disorder start with a decrease in lymphocytes, followed by other symptoms like alopecia, skin erythema, haematopoietic damage, gastrointestinal damage, and CNS damage at higher radiation doses.

In late-onset disorders, the main health effect is a higher risk of cancer and genetic defects. These are recognised as stochastic effects as they occur without the threshold. These are due to excessive exposure to doses of radiation.

Overexposure to radioactivity yields varied and numerous symptoms. The usual symptoms are

1. A decreased white blood cell count,
2. Gastrointestinal disturbances producing diarrhoea,
3. Nausea, and
4. General debility.

Overexposure may also result in loss of hair, ulceration of the mouth, unhealing ugly skin blemishes, and internal bleeding. Much more dangerous than the outward symptoms are the latent, long-term effects of exposure. These might be genetic mutations leading to cancer and even premature ageing (ironically, radiation is also employed to kill tumour cells).

It was found that ionising radiation also affects living beings at the atomic level. Whether the source of radiation is natural or artificial, whether the dose is modest or large, there will always be some biological effects associated with the radiation. When radiation enters the cells, any or all of the following events may take place:

- The radiation may traverse through without any damage,
- The radiation may harm the cell, but the cell restores the damage, or
- The radiation may cause injury to the cell, but the damage is not improved, and the cell replicates itself in the damaged form, or
- The cell dies.

Check Points

- What is the effect of radiation on humans?
- State the difference between direct and indirect effects.
- How does radiation affect different biological molecules?

Thus, it can be concluded that human exposure to radiation disturbs normal metabolic processes occurring in cells and organs by inducing gene and protein expression. Different compounds in the body are constantly being synthesised, broken down, and/or converted into different products. The dynamic nature of several aspects of metabolism can be investigated by using radiotracers that are introduced intravenously into the systemic circulation. By using gas or liquid chromatography-mass spectrometry (GC/MS or LC/MS), their fate can be traced, which will provide information on the metabolism of the tracer. Radioisotopes have been utilised in several areas of biomedical research due to the simplicity with which radioactivity can be determined and measured using relatively inexpensive and straightforward detection techniques.

Further Reading

Bajaj S, Singh A, Kalita B, Yashavarddhan MH, Ranjan R, Farooqi H, Gupta ML (2017) Gamma radiation Induced Intestinal Proteomic modulation in mice: a two Dimensional electrophoretic analysis. Defence Life Science Journal 2(3): 327-334. https://doi.org/10.14429/dlsj.2.11668.

Beck JN, Lamberty CM (2007) Thermal neutron activation analysis- An important analytical tool. 19–55. Taylor and Francis online. https://doi.org/10.1081/ASR-120004372.

CHAPTER 10—Atomic Processes. Sustainable Nuclear Power (2007) Sustainable World 249-282. https://doi.org/10.1016/B978-012370602-7/50027-2

Coomber DI (ed) (1975) Radiochemical Methods in Analysis (1975). Plenum Press New York and London. ISBN-13: 978-1-4613-4403-2. e-ISBN-13: 978-1-4613-4401-8. https://doi.org/10.1007/978-1-4613-4401-8.

Feng H, Yu Z, Chu PK (2006) Ion implantation of organisms. Materials Science and Engineering: R: Reports 54 (3–4): 49-120. Science Direct. Elsevier. https://doi.org/10.1016/j.mser.2006.11.001.

Glass WA, Varma MN (1991) Physical and Chemical Mechanisms in Molecular Radiation Biology. Basic Life Sciences Volume 58. Plenum Press. New York and London. ISBN: 978-1-4684-7627-9.

Hayes AW, Kruger CL (2014) Chapter Neurotoxicology. Hayes' Principles and Methods of Toxicology. 6th edn. Imprint CRC Press. Taylor and Francis Group. ISBN9780429099083. https://doi.org/10.1201/b17359.

Henriksen T, Maillie HD (2013) Radiation and Health. Taylor & Francis. ISBN 0-415-27162-2.

Hill DJT, Garrett RW, Ho SY, O'Donnell JH, O'Sullivan PW, Pomery PJ (1977) Radiolysis of model polypeptides in the solid state: Radiation sensitivity and side chain structure. Radiation Physics and Chemistry 17 (3): 163-171. https://doi.org/10.1016/0146-5724(81)90267-3.

Kim Y, Suh S-H, Lee I-K, Wolfe RR (2016) Applications of stable, nonradioactive isotope tracers in in vivo human metabolic research II. Experimental & Molecular Medicine 48: e203. https://doi.org/10.1038/emm.2015.97.

Krull UJ, Thompson M (2017) Analytical Chemistry. Encyclopedia of Physical Science and Technology (3rd edn.) 543-579. Science Direct Academic Press. https://doi.org/10.1016/B0-12-227410-5/00025-9.

Krishan G (2012) Livestock Nutrition: Analytical Techniques. New India Publishing Agency. ISBN: 9789380235967

Menon SS, Uppal M, Randhawa S, Cheema MS, Aghdam N, Usala RL, Ghosh SP, Cheema AK, Dritschilo A (2016) Radiation Metabolomics: Current Status and Future Directions. Frontiers in Oncology 6 (20): 1-10. https://doi.org/10.3389/fonc.2016.00020.

Meyers RA (ed) (2001) Encyclopedia Of Physical Science And Technology (3rd edn.) Artificial Intelligence. Kurfess FJ 609-629. VDOC.PUB library.

Miyoshi S, Matsuoka K, Nishimura S, Veltkamp SA (2011) Radioisotopes in Drug Research and Development: Focus on Positron Emission Tomography. In Radioisotopes Singh N (ed). https://doi.org/10.5772/22728

Organization of a Radioisotope Based Molecular Biology Laboratory (2006) International Atomic Energy Agency.

Parry SJ (2005) Activation Analysis: Neutron Activation. Encyclopedia of Analytical Science (2nd edn.) 1–10. Elsevier. https://doi.org/10.1016/B0-12-369397-7/00002-9.

Rahman MH (2019) Radiation hazard, safety, control and protection. Faridpur Med Coll J 14(2): 100- 103. https://doi.org/10.3329/fmcj.v14i2.48188.

Räisänen SJ (2005) Encyclopedia of Analytical Science (2nd edn.). 415-423. https://doi.org/10.1016/B0-12-369397-7/00583-5.

Rostampour N, Almasi T, Rostampour M, Mohammadi M, Sani KG, Khosravi HR, Pooya SMH, Golzar B, Vesal Nj (2012) An investigation of gamma background radiation in Hamadan province, IranRadiation Protection Dosimetry 152 (4): 438–443. https://doi.org/10.1093/rpd/ncs062.

Sharma A (1999) Enzyme Immunoassays: Overview. Encyclopaedia of Food Microbiology. 625-633. Elsevier. https://doi.org/10.1006/rwfm.1999.4500.

Use of Radiation in Molecular Biology

Key Learning Objectives

- Radioisotopes used in molecular biology
- Radioisotopic-based methods
- Advantages of radioisotopes in molecular biology

Overview

Since the beginning of molecular biology studies, radionuclide dependent methods have been a key part of their development and are the major tools for their use in diagnosis and study. The sensitivity, specificity, and efficiency of radionuclide-based molecular methods have been demonstrated. They are also useful in quantitative measurements. In molecular biology, the foremost important application of radioisotopes is the radioactive labelling of nucleic acids. In vivo and in vitro, they can be integrated into DNA, RNA, and protein molecules. The existence or metabolism of macromolecules can then be studied or tracked. The use of radioisotopes in various studies including DNA and RNA sequencing, DNA replication, transcription, synthesis of complementary DNA, recombinant DNA technologies, etc. has been crucial to recent developments in molecular biology that have resulted in advances in gene manipulation. The use of radioisotopes makes it possible to detect extremely small quantities, making high-sensitivity experimental methods possible.

A. Pathak, *Tools and Techniques in Radiation Biophysics*,
https://doi.org/10.1007/978-981-99-6086-6_12

In molecular biology, the most important use of radioisotopes is the radioactive labelling of nucleic acids and proteins. For the study and diagnosis of human and animal diseases, many radioisotope-based techniques are crucial.

1 Radioisotopes Used

A wide variety of radioisotopes are being used these days to study various aspects of molecular biology ranging from the behaviour of nucleic acids (DNA and RNA) and proteins and their interactions in the cells. Some examples of common isotopes utilised in the biological sciences include: along with their applications are summarised in Table 1.

In most laboratories, ^{14}C, ^{35}S, ^{32}P, and ^{125}I are used for labelling of amino acids and proteins. Except for ^{125}I, all emit β rays: ^{3}H emit low energy beta particles, ^{14}C emit medium energy beta particles and ^{35}S emit high energy beta particles.

Table 1 List of frequently used radioisotopes in molecular biology

S. No.	Radioisotope	Application
1	^{14}C	As a metabolic tracer
2	^{45}Ca	Used in the study of calcium metabolism
3	^{60}Co	Sterilisation of medical equipment and food
4	^{3}H	As a metabolic tracer
5	^{125}I	Labelling antibodies in radio immune assay and other gamma counting procedures involving *in vitro* proteins
6	^{32}P	Labelling DNA and protein
7	^{33}P	DNA sequencing
8	^{35}S	Used in biochemical research to radioactively label proteins
9	^{65}Zn	Investing transport phenomenon

2 Radioisotopic-Based Methods

A number of molecular biology techniques have developed that use radioactive isotopes. Some of these are described below:

2.1 *In Situ Hybridisation*

Key Concept
In situ hybridisation was first described in 1969 for localising specific DNA sequences on chromosomes in cytological preparations of Xenopus oocytes.

In situ hybridisation is a very powerful technique in research and diagnosis. This technique allows precise localisation of a specific segment of nucleic acid in a histological section by using complementary DNA or RNA or modified nucleic acid strand (i.e., probe)—can be radioactive as well as nonradioactive. There are four different types of probes that can be used in *in situ* hybridisation: double-stranded DNA probes, single-stranded DNA probes, RNA probes (ribo probes), and synthetic oligonucleotides. Radioactive labelling commonly involves the use of ^{3}H, ^{32}P, or ^{35}S, to the nucleotides. However, sometimes ^{14}C and ^{125}I can also be used.

For mapping brief low copy number sequences of DNA without using isotopes, PCR *in situ* is another option to *in situ* hybridisation. This technique uses a form of PCR on fixed metaphase chromosomes on microscope slides, with the temperature cycling controlled by a flat bed thermal cycler. Further development of this technique is expected to achieve the target of localising unique sequences. However, radioactive labelling has a number of benefits. Radioisotopes can be readily incorporated into probes using various enzymes, radioactively labelled probes are usually more sensitive as compared to non-radioactively labelled probes, and radioactive probes enable quantitative analysis. Immunohistochemistry also allows researchers to discover how the distribution of particular nucleic acids relates to the formation of protein products corresponding to the target gene and also their relationship to cellular structures. *In situ* hybridisation has difficulty in identifying targets with minimal copies of DNA and RNA.

2.2 *Dot Blot Assay*

This is a probe based rapid screening molecular biology technique that facilitates detection of proteins in different samples. It is based on the homology of two strands of DNA (i.e., DNA:DNA or DNA:RNA or RNA:RNA). A reporter molecule is attached to a probe that is a single-stranded nucleic acid (RNA or DNA) in this technique. After isolating nucleic acid from the test sample, the labelled probe is then allowed to form a hybrid with it. Dot blots are nearly similar to Western blots as they also involve the application of a small quantity of sample directly on a membrane, which is then examined using a specific probe of antibody or nucleic acid to detect target molecule. As opposed to western blotting, they don't need the protein to be separated electrophoretically on a gel.

Dot blots are useful to estimate the protein content in unrefined preparations like tissue culture supernatant or ascites; to determine the effectiveness of antibody-based detection system; or to determine the appropriate concentration of antibody for Western blotting. However, dot blots are not used to calculate the molecular mass of a protein or to distinguish between distinct protein forms because there is no protein separation step (for example, cleaved proteins or phosphorylated proteins).

The labels used here can be either radioactive or not. Autoradiography is utilised for detection if radioactive probes are used. The discovery of drug-resistant mutations in mycobacterium tuberculosis is extremely common using radioisotope dot blot hybridisation. Due to its combination of precision and resilience, this method is also excellent for widespread surveillance of medication resistance in the malarial parasite. ^{32}P- based dot blot hybridisation is also used for the detection of hepatitis C virus easily and with high accuracy and it provides high intensity dots so that the results are easily interpreted.

2.3 Restriction Fragment Length Polymorphism (RFLP) Analysis

Another method used in molecular biology is RFLP, which takes advantage of variations in DNA sequences that are closely related. In restriction fragment length polymorphism analysis (RFLP), specific restriction endonucleases recognise and cleave DNA molecules at certain places, resulting in fragments of amplified DNA. Southern blotting is used to transfer the DNA fragments to a membrane after separation by agarose gel electrophoresis. The complementarity of the fragments to the probe is then determined by hybridising the membrane with a tagged DNA probe. Single, low-copy genomic DNA or cDNA clones are used as RFLP probes. A labelled DNA sequence is used as an RFLP probe that combines the digested DNA fragments after they were separated by agarose gel electrophoresis, thereby revealing a distinctive blotting pattern of a distinct genotype at a specific locus. These complimentary probes are also radioactively labelled, allowing them to be seen as bands using autoradiography.

Analysis of RFLP genomic diversity aids in the mapping of the individual genome and the study of hereditary diseases. Genome mapping aids in the examination of distinctive genetic patterns for organism identification and differentiation. It is also useful for calculating the frequency of recombination between restriction sites. Genetic disease analysis deals with the analysis of the gene among other family members after it was identified for particular genetic or hereditary disease. RFLP analysis is regarded as the basis for the first genetic fingerprinting techniques, which were helpful in identifying samples obtained from crime scenes, establishing paternity, and describing genetic diversity or breeding trends in animal populations. The epidemiology of tuberculosis has benefited greatly from this method. Mycobacterium tuberculosis may be reliably and repeatedly typed using this technique.

2.4 Single Stranded Conformational Polymorphism (SSCP)

This is a common, sensitive, and straightforward method for mutation detection and genotyping. It is the post- PCR technique which is useful in screening for mutations that are randomly dispersed throughout the exons and are not restricted to a specific hot spot. After amplification of the target nucleic acid by PCR, the radiolabelled PCR products are highly diluted, denatured to a single strand with high temperature and exposure to other denaturing conditions. After that, they were separated by polyacrylamide gel electrophoresis. The bands were then visualised by autoradiography. It is a very sensitive method and is also called radio-labelled "hot" SSCP. During the procedure, the gel is run at low voltage to avoid temperature induced conformational changes. This means accurate control of the temperature of the gel is very important to get consistent results. If a mutation is present in the amplified sequence, that will alter the folding of the sequence and, consequently, its electrophoretic mobility. Consequently, the wild-type and mutant sequences will move differentially in the gel. Bands that appear at positions other than those seen in the wild-type strain are indicative of mutations. Consequently, SSCP analysis determines sequence variations (i.e., single-point mutations, small deletions, and insertions) via differences in their electrophoretic mobility. The main benefit of the SSCP technique is its responsiveness in detecting mutations (substitutions, minor deletions, and insertions) at various positions in DNA fragments. However, it is unable to detect the accurate nucleotide change. This necessitates the use of a supplementary technique, like DNA sequencing. Second, since SSCP bands cannot be seen directly, it is challenging to excise them from the gels for sequence depiction.

2.5 Amplified Fragment Length Polymorphism (AFLP)

This technique is based on the idea of selectively amplifying a subset of restriction fragments from a complicated mixture of DNA fragments generated after genomic DNA was digested with restriction endonuclease. AFLP is another PCR-based instrument used in genetic experiments, DNA fingerprinting, and genetic engineering. Only a minimal amount of genomic DNA is required. This DNA is digested with a set of restriction enzymes—one having an average cutting frequency (such as EcoRI) and the other with a high cutting frequency (like MseI or TaqI). Double-stranded oligonucleotide adapters are synthesised so that one end crosses the cutting site and is complimentary to it, while the other end continues into the unknown sequence next to the restriction fragment. Adaptors are then ligated to the sticky extremities of the restriction fragments. Adaptor-specific primers are used for PCR amplifications that is radiolabelled. Typically, ^{32}P-labelled AFLP primers are utilised. After performing polyacrylamide gel electrophoresis and autoradiography, a highly useful pattern in the form of bands is obtained. Polymorphic patterns are obtained from various strains because they result from alterations in the sequences of the restriction sites, in the sequences adjacent to the restriction sites, and insertions and deletions in the amplified fragments. AFLP is a highly sensitive technique for detecting DNA polymorphisms. Radioisotopes can be used to detect this polymorphism.

Various microbes have been characterised using this method. Acinetobacter, Bacillus anthracis, Legionella, Salmonella enteritidis, and Chlamydia sp. are some of these (genital, ocular, and respiratory pathogens). It is also a beneficial option to micro satellite analysis of parasite genetic investigation. AFLP technology can simultaneously detect multiple polymorphisms in distinct genomic regions. The technique is also exceedingly sensitive and reproducible. Therefore, AFLP is now frequently used to identify genetic variations in closely related species of plants, fungi, animals, and microorganisms. This technology is also used in forensic and paternity tests to identify minute differences within populations, as well as in genetic linkage studies to create maps for quantitative trait locus (QTL) analysis.

2.6 *Mismatch Cleavage Assay*

This method can detect point mutations and also single base pair deletions in 1.5–1.7 kB sequences. DNA is extracted and gene of interest is amplified using PCR. It is hybridised with radiolabelled RNA probes acquired from the familiar wild sequences. These DNA-RNA heteroduplexes having a varying number of mismatched bases are treated with RNase A, that cuts wherever a mutation causes a mismatch with the original sequence. The cleavage-resistant products (RNA that is resistant to RNase) are then analysed by denaturing polyacrylamide gel electrophoresis and then autoradiography. For each strain, a distinctive pattern of cleavage-proof RNA fragments is produced, which can be used to identify the isolate. This technique is suitable for promptly analysing several genetic variants. It is a screening procedure for presequencing and for detecting unknown mutations. This assay has been used in the genetic investigation of cancer, such as p53 mutants in colorectal cancer, and circulating variants of viruses, like the human respiratory syncytial virus, a frequent reason of acute respiratory tract infection in children less than 5 years old. Such an analysis of viral variants is crucial for tracking viral evolution and generating basic data for vaccine development.

2.7 Heteroduplex Tracking Assay (HTA)

It detects sequence differences between normal DNA and the DNA being tested. It is frequently used as a screening method to detect point mutations in a gene, for instance, this assay is used to determine the genetic relationship between two or more viral genomes. It is also utilised for the genotyping of HCV. This method is more precise than RFLP for subtyping HCV, has the potential to identify new variants, and is valuable for epidemiological studies. The heteroduplex monitoring assay can detect single, specific mutations or linked mutations HTA permits the detection of HIV populations with a small representation that can be quantified on the basis of the band intensity obtained from autoradiography. In this assay, a radioactively (^{35}S) tagged probe binds to a PCR product produced from the total virus (e.g., HIV) population in an individual to form probe-PCR product heteroduplexes containing a varying number of mismatches that can be separated by gel electrophoresis. Clustered mutations, insertions, or deletions lead to altered heteroduplex movement, that can show definite subsets of viral genomes. The assay is quantitative and sensitive-thanks to the use of a radiolabelled probe. In conjunction with denaturing gradient gel electrophoresis (DGGE), it is a sensitive and readily readable test for detecting single base pair substitutions, such as those found in drug-resistant mutants. HTAs rely on mobility shifts in the gel to locate single nucleotide polymorphisms and are sensitive to minority forms. They also assist in determining the relative frequency of sensitive and resistant parasite forms in a patient.

2.8 DNA Sequencing

Key Concept
Maxam–Gilbert sequencing is a method of DNA sequencing developed by Allan Maxum and Walter Gilbert in 1976–1977. This is also called Gilbert Chemical-Degradation Sequencing.

DNA sequencing is the process of figuring out the exact sequence of nucleotides that make up a DNA strand. It is a benchmark for finding mutations. There are several ways available, including the Sanger (i.e., enzymatic) method, the Maxam and Gilbert (i.e., chemical) method, and emerging methods like pyrosequencing in real time by the measuring of released pyrophosphate (PPi). The Maxam–Gilbert approach necessitates radioactive labelling of DNA at one end and purification of the DNA fragment that will be sequenced. Chemical treatment is done to produce breaks thus generating a series of labelled fragments. These fragments are separated by size in gel electrophoresis and then visualised using autoradiography from which the sequence can be inferred. DNA sequencing has been an important tool in Human

Genome Project. It can be used to diagnose inherited disorders, focus on cancer treatments and recognise health risks.

The emergence of quick DNA sequencing technologies has greatly accelerated the pace of research and discovery in biology and medicine and many applied disciplines like diagnosis, biotechnology, forensic science, virology, and systematics. Sequencing is also useful to analyse genomes and proteins they encode. Sequencing data enables the identification of maladies and phenotypes, as well as potential dry targets.

2.9 DNA "Foot Printing" for Analysis of Protein-Nucleic Acid Interactions

Regulatory proteins and DNA interactions control gene expression. Transcriptional regulatory proteins anneal to particular DNA target sequences (called promoters) in the particular gene. Several *in vitro* and *in vivo* methods are used to discover these proteins by finding their capability to bind to DNA and locating their binding site of the gene. This allows for the development of therapeutic strategies that downregulate the production of these proteins to obstruct these pathways and hinder infections or cancer cells. DNA foot printing is a technique used to identify protein-binding regions in DNA molecules. Examples include identifying regions of binding of transcriptional factors in a gene and initiate transcriptions, and identifying hormone response elements (HREs), which are distinct sequences of attachment of the hormone receptor complexes. DNA foot printing is studied on the assumption that if a protein binds to a target DNA, that is secured from cleavage by compounds or enzymes (e.g., DNAse I). This DNA is radio labelled at one end (to enable visualisation by autoradiography) and is incubated with the nuclear extract so that the proteins bind to DNA sequences they recognise. Deoxyribonuclease I then partially degrades the DNA. If the protein binds to the DNA and protects a portion of DNA from cleavage, voids (cleavage sites) are detected in the ladder generated by DNase I digestion. These sites correspond to the binding site. It suggests that the protein may function as a transcriptional regulator by binding to DNA.

There is no radiolabelled band at the site of binding of protein on the autoradiogram. Foot printing can be done either by *in vitro* method or *in vivo* method. *In vivo foot printing* enables the monitoring of alterations in DNA- protein binding that occur within a living cell. For this dimethyl sulphate (DMS) is used because it can readily traverse the cell membrane of live cells. As a cleavage agent, DNase I or ultraviolet light can be used. After the DNA has been cut, the cells are disintegrated and DNA is purified. Ligation-mediated PCR serves an another option to *in vivo* foot printing. After the genomic DNA is treated with a cleavage agent and single strand break occurs, the DNA is isolated and then a linker is attached at the break points. A region of interest is amplified using a gene-specific primer, and shows a foot print of protein binding site on a polyacrylamide gel.

Protein specificity can be evaluated at numerous genomic locations using *in vivo* foot printing and immune precipitation. The DNA binds to a protein of interest, immunoprecipitated with an antibody against that protein, and then specific region binding are evaluated using the DNA foot printing technique.

The binding strength of a protein to a section of DNA can be evaluated using a method called quantitative foot printing, that is a slight modification of DNA foot printing. The binding affinity of a protein can be measured by using a foot printing experiment in which the protein is present at varied concentrations and the presence of the footprint is examined with an increase in the concentrations.

2.10 Scintillation Proximity Assay (SPA)

In order to quickly and accurately evaluate a wide range of molecular interactions in a homogenous solution, the scintillation proximity assay (SPA) was commercially released in 1991. It makes use of tiny beads that contain a scintillator that, when excited, emits light. This stimulating event occurs only when radiotagged compound of interest is in close proximity to the bead's surface. ^{3}H and ^{33}P are the radioisotopes employed. The SPA method relies on the conversion of radioactive decay into energy, which emits light photons that may be detected by using equipment like photo multiplier tubes in scintillation counters or CCD pictures. In fields that require the detection and quantification of radioactivity, this method is widely used. Receptor-ligand binding studies, enzyme assays, radioimmunoassays, protein-protein and protein–DNA interaction studies are all possible using the SPA system. The pharmaceutical sector makes extensive use of it for medication research and design.

2.11 Microarray Chip Technology

Microarrays is a collection of high- density microscopic single-stranded DNA attached to a solid surface. DNA microarrays are used by scientists to genotype numerous areas of a genome rapidly and effectively or to evaluate the expression levels of several genes at once. A computerised image-analysis system, such as a phosphorimager, is used to read the data. Smaller sample sizes, reduced reagent use, quicker reaction times, and more data generated from a single analysis are all benefits of arrays. Array technology can be used for large-scale DNA genotyping and detection of SNP, as well as for studying DNA-protein interactions, such as the of DNA binding proteins identification that are involved in cellular signalling and oncogenesis.

DNA microarray technology aids in identifying new genes, as well as knowledge of their function and expression levels under various conditions. This also enables researchers to learn more about a variety of diseases, including cardiovascular diseases, mental disorders, infectious disease, and cancer in particular at the molecular level. Microarray technology is widely used in pharmacogenomics, which is the study of association between therapeutic drug responses and a patient's genetic profile.

Chips with up to 64,000 immobilised sensors can now be produced by commercial manufacturers. Chips are commercially available for HIV typing (i.e., reverse transcriptase and protease genes), cytochrome P450 gene, and p53 gene, among others. Nonetheless, this cutting-edge technology is quite expensive, with sophisticated expression profiling chips (which are not even reusable) costing thousands of dollars. Isotopic microarrays with several hundred probes immobilised on nylon wafers are suitable for radioisotopic labelling and detection. These are a relatively inexpensive option for automation with a lower throughput, making them more suitable for laboratory contexts. In addition, the required quantity of sample (target DNA or RNA) is only in the nanogram range, which is nearly 100-fold less than for non-radioactive high throughput microarrays. This new microarray method can be used for identifying microorganisms which consume specific compounds, for example, ^{14}C-labelled substrate within complex microbial communities. The identification can be performed using FISH and microautoradiography. A *protein microarray* is a high yield method to monitor the activities and interactions of proteins, as well as to determine their function on a large scale. Its chief advantage is the ability to track a large number of proteins simultaneously.

2.12 Isotope Coded Affinity Tags (ICAT)

In proteomics, two-dimensional gel electrophoresis (2-DE) mass spectrometric analysis of the separated and enzymatically digested protein. is the standard protein separation method. The chief disadvantage of 2D electrophoresis is that it is labour- and time-intensive. Quantification issues also exist. Numerous drawbacks of two-dimensional gel electrophoresis have been surmounted by the automated 2DE-MS, the use of multidimensional liquid chromatography (LC), and tandem mass spectrometry (MS/MS). Isotope-coded affinity-tag (ICAT) peptide labelling is a nowadays potent alternative for proteomic analysis. This is an isotopically labelling technique for quantitative proteomics that employs chemical labelling reagents and is detected by mass spectrometry. The chemical probes utilised in this study consist of three components: a reactive group for labelling of amino acid side chain (e.g., iodoacetamide to modify cysteine residues), an isotopically coded linkage, and a tag (e.g., biotin) for affinity isolation of labelled proteins/peptides. Isotope labels (e.g., ^{18}O, ^{15}N) are incorporated into the protein being analysed. Liquid chromatography is used to separate the tagged peptides, and mass spectrometry is used to analyse them. Using stable isotope coded affinity tags, large-scale quantitative analysis of complex protein mixtures is conceivable. This provides an ideal method for accurately quantifying low copy number proteins that play a crucial function at the cellular or tissue level.

One sample is labelled with the isotopically light (d0) probe and the other with the isotopically heavy (d8) version for the quantitative comparison of two proteomes. ICAT is a highly automated method that enables tandem MS to sequence peptides directly. ICAT's preference for cysteine-rich proteins, however, is its primary limitation. The ICAT reagent frequently interferes with MS fragmentation and lowers the quality of the derived MS spectra. This limitation has, however, been successfully surmounted through the development of a cleavable ICAT reagent that is removed prior to MS fragmentation and analysis. Another disadvantage is that the reagents used in ICAT only label cysteine- containing peptides. Therefore, if the particular protein does not contain any cysteine residue, it will not be detected.

3 Advantages of Radioisotopic Methods

Though a lot of new alternative methods have developed but still radioisotopes are preferred for several applications because they offer numerous advantages in molecular biology and biochemical analysis as compared with non- radioisotopic experimental procedures.

3.1 *Sensitivity*

Using a scintillation counter, X-ray film (autoradiography), or imaging plate, trace amounts of radioisotopes can be readily detected. Due to their exceptional sensitivity, which enables the detection of as little as 0.1 pg of target DNA, probes labelled with isotopes (e.g., ^{32}P) play a significant role in DNA hybridisation, a widely employed molecular biology technique. When compared to ethidium bromide staining, the sensitivity of isotope-labelled products can increase by up to 125 times. In addition to the radioisotopic method, non-radioisotopic methods have also been used to label DNA with digoxigenin (DIG). The radioisotopic technique, however, is more sensitive than the DIG method.

3.2 *Cost Effectiveness*

Radioisotopic hybridisation is less expensive than non-radio isotopic procedures, according to cost estimates. A single glass "chip" with immobilised probes and fluorescence-based detection can be used to detect thousands of genes using microarrays. Due to the exorbitant cost of equipment, this technology is still out of reach for academic laboratories. In such situations, a relatively low-cost method employing nylon microarray devices and radio isotopic (autoradiography) detection is a viable alternative.

3.3 High Accuracy

Use of radioisotopes like ^{32}P can provide highly accurate, clear and quantitative results in studying various processes in the cell. For example, protein phosphorylation studies in the cell. Here, particular antibodies are made and used. The precision and sensitivity of a test are contingent on the quality and specificity of the antibodies used. During such studies using antibodies, sometimes false signals also appear. However, labelling proteins with radioactive -^{32}P ATP in cell-free systems (*in vitro*) or in experiments involving cells, tissues, or organisms (*in vivo*) yields accurate and precise results.

3.4 Reflection of Natural Conditions

In most cases, the radioisotope-tagged organic and inorganic compounds have the same chemical properties as normal compounds and elements, so they can be used to track compounds and elements in cells, tissues, and organisms without artificial conditions.

Check Points

- Name the various radioisotopes used for studying different molecular biology techniques.
- What is *in situ* hybridisation?
- What is the importance of the dot blot assay?
- Write a note on RFLP.
- What is the "hot" SSCP method?
- How are radioisotopes used in AFLP?
- Write about the advantages of DNA sequencing.
- What are DNA foot printing techniques used for?
- What are microarrays? Explain DNA microarray technology.

Thus, it can be seen that the gene probes are usually labelled with radioactivity tagged with fluorescent material. The radioactive tag allows the visualisation of the DNA binding to the cell or a chromosome within the sample. Labelling can be done at either 3′ or the 5′ end. ^{32}P has traditionally been the preferred isotope, but ^{35}S has also been used successfully. When high resolution (as in *in situ* hybridisation) or extended probe stability are required, ^{35}S is particularly useful. A hybridisation probe is a fragment of DNA or RNA that is typically 15–10,000 nucleotides long and can be radioactively or fluorescently labelled in molecular biology. It can be utilised to identify the presence of complementary nucleotide sequences in RNA or DNA that have been analysed. The labelled probe is first denatured into single-stranded DNA and then hybridised to the target single-stranded DNA (Southern blotting) or nucleic acid (Northern blotting) immobilised on a membrane or *in situ*. The DNA probes can be labelled in a variety of ways, including nick translation, random primed labelling, and PCR. Each fragment length is regarded as an allele and is suitable for genetic analysis. *In vitro* transcription labels the RNA probes.

To detect hybridisation of a probe to its target sequence, the probe is labelled with a radioactive or fluorescent molecular marker. For radioactive detection, ^{32}P is incorporated into the phosphodiester bond of the probe DNA. These are then detected by visualising the hybridised probe via autoradiography or by GM counter. Nucleic acid probes can also be labelled using other radioactive probes like ^{3}H, ^{125}I, and ^{35}S. Using radiolabel is sensitive, reliable and predictable of detection. Due to safety concerns, cost, and the disposal of radioactive waste products, radiolabelled DNA probes are no longer as prevalent as they once were. Handling dangerous radioisotopes also necessitates special and stringent laboratory procedures. In addition, attenuation or decay of isotopes is inevitable, which may affect experimental repeatability; as a result, the radiolabelled method necessitates frequent isotope repurchasing, substrate relabelling, and radioactive waste disposal. However, radiolabelled probes are the most sensitive, as they offer the maximum level of resolution currently available in hybridisation assays. High sensitivity implies that low concentrations of a probe target hybrid can be detected; for instance, ^{32}P- labelled probes can detect single- copy genes in only 0.5 g of DNA.

Science and medicine have been revolutionised by molecular biology because it can explain how several intracellular phenomena occur. Various molecular biology techniques e.g. dot blot assay, restriction fragment length polymorphism (RFLP), single-stranded conformational polymorphism (SSCP), amplified fragment length polymorphism (AFLP), mismatch cleavage assay, heteroduplex tracking assay (HTA), DNA sequencing, microsatellite detection, scintillation proximity assay (SPA), microarray chip technology, isotope coded affinity tags (ICAT), etc. may use isotopes for detection purposes.

When using radioisotopes in molecular biology, the most important considerations are their safe handling and appropriate disposal. However, both of these issues can be resolved quickly and easily with the proper training, the use of protective equipment, and adherence to the guidelines for the use and disposal of radioisotopes. In addition, the safety of low energy isotopes, which are typically used in molecular biology techniques, is less of a concern because these isotopes emit significantly weakened particles.

Further Reading

Amsath A (2008) Textbook of Basic Biotechnology 2nd edn. M. M. A. Publications.

Barakat SM (ed) (2012) Salmonella—A Dangerous Foodborne Pathogen. Published by InTech. ISBN: 978-953-307-782-6.

Dreyer M, Erdmann J, Rehmann-Sutter C (eds) (2016) Genetic Transparency? Ethical and Social Implications of Next Generation Human Genomics and Genetic Medicine. Series. Life Sciences, Ethics and Democracy, Volume: 2. ISBN: 978-90-04-31189-3. Brill Publishers.

Esh AMH. (2014) Molecular Markers and Phytopathology Chapter 15. Biological Controls for Preventing Food Deterioration. 341–376. Sharma N (ed). https://doi.org/10.1002/9781118533024.ch15.

Hunter Tc, Andon NL, Koller A, Yates III JR, Haynes PA (2002) The functional proteomics toolbox: methods and applications. Journal of Chromatography B 782 (1–2): 165-181. https://doi.org/10.1016/S1570-0232(02)00570-6.

Jensen E (2014) Technical Review: An in situ hybridisation. The Anatomical Record. Wiley Periodicals, Inc. 297:1349–1353.

Kagan J, Srivastava S (2005) Mitochondria as a target for early detection and diagnosis of cancer. Critical Reviews in Clinical Laboratory Sciences 42 (5-6): 453–472. https://doi.org/10.1080/10408360500295477.

Kawachi M, Nagasaki-Takeuchi N, Kato M, Maeshima M (2011) Application of radioisotopes in biochemical analyses: Metal binding proteins and metal transporters. 115-126. In Radioisotopes—Application in Bio- Medical Science. Singh N (ed). ISBN 978-953-307-748-2.

Kumar A, Misra P, Dube A (2013) Amplified fragment length polymorphism: an adept technique for genome mapping, genetic differentiation, and intraspecific variation in protozoan parasites. Parasitology Research 112: 457–466.

Lee Y-C, Moon J-Y (2020) 171–197. Springer Publications. https://doi.org/10.1007/978-981-15-1293-3_10.

Luo W, Yu Y, Wang R, He D, Wang C, Zeng X, Chen X, Tan X, Huang T, Wu X (2013) P7 peptides targeting bFGF sensitize colorectal cancer cells to CPT-11. International Journal of Molecular

Manosas M, Camunas-Soler J, Croquette V, Ritort F (2017) Single molecule high-throughput footprinting of small and large DNA ligands. Nat Commun 304. https://doi.org/10.1038/s41467-017-00379-w.

Morris CM (2004) Functional genomics and proteomics: application in neurosciences J Neurol Neurosurg Psychiatry 75:529–538. https://doi.org/10.1136/jnnp.2003.026260.

Organization of a Radioisotope Based Molecular Biology Laboratory (2006) International Atomic Energy Agency.

Ravi I, Baunthiyal M, Saxena J (eds) (2014) Advances in Biotechnology. Springer New Delhi Heidelberg New York Dordrecht London. ISBN 978-81-322-1553-0. https://doi.org/10.1007/978-81-322-1554-7.

Resch W, Parkin N, Stuelke EL, Swanstrom R (2000) A multiple-site-specific heteroduplex tracking assay as a tool for the study of viral population dynamics. 98 (1): 176-181. https://doi.org/10.1073/pnas.98.1.176.

Singh N (ed) (2011) Radioisotopes—Applications in Bio- Medical Science. IntechOpen. ISBN 978-953-307-748-2. eBook (PDF) ISBN 978-953-51-6576-7.

Walker JM, Rapley R (eds) (2008) Molecular Biomethods Handbook. (2nd edn.) Humana Press. ISBN: 978-1-60327-375-6.

Wilson K, Walker J (eds) (2005) Principles and Techniques of Biochemistry and Molecular Biology. 6th edn. Canbridge University Press. ISBN 978-0-521-82889-5 (hardback). ISBN 978-0-521-53581- 6 (paperback).

Wilson KE, Ryan MM, Prime JE, Pashby DP, Orange PR, O'Beirne G, Whateley JG, Bahn S, Zolla L, D'Alessandro A (2011) Chromatography of Membrane Proteins and Lipoproteins Peptides and Proteins. Encyclopedia of Analytical Chemistry: Applications, Theory, and Instrumentation. Wiley Online Library. https://doi.org/10.1002/9780470027318.a1607.pub2.

Other Applications of Radiations

Key Learning Objectives
- Industrial applications of radiations
- Applications of radiation in agriculture
- Ecological studies with radioisotopes
- Nuclear power plants

Overview
In today's world, radiation is being used in almost all fields: medicine, academics, industry, generating electricity, agriculture, archaeology (carbon dating), space exploration, law enforcement, geology, etc. It has already been discussed in detail in the previous chapters that radioisotopes have several applications in medicine, diagnosis, and therapy. Many applications use a special technique called the "tracer technique". Some other applications of radioisotopes are discussed in this chapter.

1 Industrial Applications

Radioisotopes are useful in many industrial processes. These are useful tools in industry, both as process monitors and radioactive tracers. Through non-destructive, sensitive testing, and effective in-process control, the radioisotopes used in industry

A. Pathak, *Tools and Techniques in Radiation Biophysics*,
https://doi.org/10.1007/978-981-99-6086-6_13

produce high-quality products while simultaneously reducing manufacturing costs. Industry uses radioisotopes in a number of ways to increase productivity and obtain information that is not possible otherwise. In mineral analysis, industrial radiography, gauging applications, and other industrial applications, sealed radioactive sources are used.

Industrial applications of radioisotopes are mainly classified into three types:

(a) Radiation processing.
(b) Non-destructive testing.
(c) Radiotracer application.

1.1 Radiation Processing

This involves the controlled use of energy from ionising radiation (i.e., gamma rays, accelerated electrons, and X-rays) to have the intended impact on the product. There are several benefits to radiation processing:

1.1.1 Radiation Sterilisation of Medical Products

Ionising radiation has been effectively used as a bactericide for commercial sterilisation of medical supplies and equipment. A dose of 25 kGy has been found to sterilise the products in the sealed containers. The gamma rays pass through this tightly sealed packaging and eradicate microorganisms that may be present in the product. Some of the products that are regularly sterilised with gamma rays are blades, bandages, syringes, sutures, absorbent cotton, wool, infusion sets, surgical instruments, and certain antibiotics and ophthalmic ointments. These medical equipments are sterilised without being exposed to toxic chemicals or extreme heat.

1.1.2 Leak Detection

Leaks in long and concealed pipelines and industrial facilities can be monitored by adding a small amount of a specific short-lived isotope, such as ^{82}Br, to the fluid. The leak in underground water can also be identified by ^{3}H or ^{24}Na. This involves separating the parts of the pipeline, administering the tracer, maintaining pressure on the liquid column, forcing liquid into the pipeline, and examining the surface for leaked radioactivity. The progress of the tracer through the pipe is tracked by a radiation detector outside the pipe. The leak or blockage is discovered when the amount of radiation detected decreases. The detection of even the smallest leakage has resulted in significant cost, energy, and labour savings. Similarly, radioisotopes can be used to detect leaks in dams and reservoirs.

1.1.3 Chemical Reactor

Chemical reactors are containers made to hold chemical reactions. For maintaining the chemical industry's production efficiency and product quality, the mixing time and mean residence time of components in process containers are crucial. The mixing rate and residence duration are calculated by adding a small amount of the radiotracer at the input location and tracking it over time. Such measurements can be done both in-stream and by sampling. These analyses are often used in the chemical and cement industries. This is accomplished by using ^{24}Na and ^{60}Co.

1.1.4 Silt Movement in Harbours

Dredging operations are performed in harbours on a regular basis to keep shipping channels clear. To choose suitable sites for depositing dredged silt and to assess the relevance of the orientation of any shipping channel in a new harbour, it is necessary to examine the silt's underwater movement pattern. A radiotracer, such as ^{46}Sc, is combined with sediment and deposited on the seabed. Using underwater radiation detectors, the radioactivity of ^{46}Sc is measured to track the movement of sediment. Such techniques are being used by the port authorities.

1.1.5 Movement of Pollution Offshore

Radiotracers are also used to investigate the pattern of waste dispersion in water bodies. It is possible to reliably determine the optimal discharge distance and timing. Such research would aid in ensuring secure and acceptable waste disposal methods and enhance environmental protection measures. For this, ^{82}Br as NH_4Br can be used as a tracer.

1.1.6 Sludge Hygiene

Sewage sludge is abundant in organic matter, nutrients, and trace elements and should be recycled for economic reasons. The unprocessed sludge contains a large number of pathogens that cause infectious diseases such as typhoid, cholera, and hepatitis. The pathogen load of sewage sludge restricts its reutilisation, though it is good nourishment for farmland. Therefore, there is a need to include a step in the treatment process to effectively de-activate the pathogenic organism with a high level of trustability and harness its beneficial nutrient properties. Appropriate treatment and disposal of sewage sludge is also important to protect the environment, the health of the citizens, the aesthetics of the cities, and the optimal use of land area. As a result, the sludge must be processed before it can be used as fertiliser. It is known that radiation can cause lethal damage to microorganisms. The gamma rays of ^{60}Co, produced artificially in nuclear reactors, are energetic enough to inactivate pathogens with an elevated degree of dependability in a decent and coherent manner. Actually, the ionising radiation interacts directly or indirectly with DNA and proteins to bring about changes which cause cell death. The pathogen concentration in sewage sludge has been found to be reduced by 3–5 kGy of radiation treatment.

1.1.7 Treatment of Flue Gases

Flue gas is the combustion exhaust gas produced by power facilities, and it may lead to environmental degradation. It contains nitrogen gas, carbon dioxide, water vapour, and a minute amount of several pollutants, including particulate matter (like soot), carbon monoxide, nitrogen oxide, and sulphur oxide. A technique called *electron beam processing* is available to transform the gaseous pollutants to the components of fertiliser, such as ammonium sulphate and ammonium nitrate. During this treatment, before being admitted into an irradiation chamber where ammonia and the oxides are transformed to their salts, the broiler's flue gas is filtered mechanically and saturated with water vapour. Actually, high-energy radiation ionises and stimulates molecules and atoms, resulting in the formation of highly reactive free radicals, ions, and molecules in excited states. These can react amongst themselves, with the oxygen and water already present in the system, or with certain chemical components added to the system intentionally, giving products that can be pipetted and collected by convenient means.

1.1.8 Radiation in Manufacturing Process

Radiation from a radioisotope is either scattered or absorbed as it travels through materials. The quantity of light that is entirely absorbed by an object is proportional to its thickness and density. Because of this feature, radiation can be used to produce goods of higher quality, such as treated leather, polyethylene foam, cables, and semiconductors. The radiation dose required varies widely depending on the type of treatment. This is also often used for the vulcanisation of natural rubber latex instead of following the conventional method that uses sulphur, which causes rubber to produce carcinogens. The operating temperature range of polymers is increased via radiation cross-linking, which gives them dimensional stability.

1.1.9 Gauging

Radioisotope gauging is based on the principle that matter located between a radioisotope and a detector will reduce the intensity of the radiation emitted by the radioisotope. On one side is a sealed source of a suitable radioisotope, and on the other is the detector. Any material's attenuation of radiation passing through it is proportional to its mass. This reduction can be used to determine the presence or absence of the material as well as the amount of material between the source and detector. This property can also be used to measure and monitor the thickness of manufactured sheet materials such as metal and plastic films and sheets, textiles, aluminium foil, newspaper, copier paper, plastics, and photographic film. To screen out fragile, thin eggs before they are packaged in egg cartons, radioactive material is also used in gauges that measure the thickness of egg shells. The same method can be used to determine whether sealed cigarette packets are complete or whether one or more cigarettes are missing. The packets are set on a conveyor belt that runs between a radioisotope and a counter. Due to less radiation absorption than with a fully filled packet, an unfilled or partially filled packet has a higher counting rate. The increase in counting rate can be converted into an electronic signal that dislodges the incomplete package from the conveyor belt.

In all industries where the concentrations of gases, liquids, and substances must be measured, gauges with radioactive (typically gamma) sources are widely used. Using radioisotope gauges kept outside the tanks, the levels of liquids in closed or large vessels can be conveniently measured and controlled. Engineers also use radioactive substance-containing gauges to measure the fluid levels of oil and chemical containers, as well as the moisture and density of soils and materials at construction sites. Controlling the filling of gas cylinders and detergent powder containers uses a similar concept. These gauges are most useful in situations where heat, pressure, or corrosive substances, such as molten glass or molten metal, make it impossible or challenging to use direct contact gauges.

Gauges are also used to measure ordinary items like the quantity of adhesive on a postage stamp, the amount of liquid in canned goods, the amount of air whipped into ice cream, and the amount of tobacco in a cigarette. The machine-tools business can also benefit from the use of radioactive tracers to monitor the degree of wear on cutting tools and drills.

Toxic pollutants, such as those produced by coal-fired power plants and industry, are also removed using radioisotopes. Electron beam radiation can be used, for instance, to clean the air of harmful gases like sulphur dioxide and nitrogen oxide.

Many of the fabrics used to create our clothing have been irradiated before being treated with a wrinkle- resistant chemical. This treatment causes the chemicals to bind to the fabric, keeping the apparel fresh and wrinkle-free without rendering it radioactive. Similarly, gamma rays are used to treat nonstick cookware to prevent food from adhering to the metal surface.

To better monitor the quality of the oil produced, the radioactive isotope ^{41}Ar is regularly introduced into the catalytic cracking towers.

In order to monitor industrial operations, radioactive sources are often used. Large wagons carrying raw materials, for instance, are passed in front of a neutron source in the cement business. The sample's nuclei take in the neutrons and then radiate their excess energy as gamma rays. The discovery of these rays enables better knowledge of the sample's composition, which can then be adjusted to meet factory standards.

Although significantly larger doses of radiation are used, radioactivity is also used to alter a material's mechanical and chemical properties. Soft wood is turned into panelling that is as durable as marble through an industrial process. Places with lots of foot traffic, like airports, public buildings, department stores, and museums, benefit from the durability of parquet flooring.

1.2 Non-Destructive Testing

This is a form of quality assurance testing that does not involve the destruction of the product or substance being tested. To find fatigue effects, structural faults, and other similar issues, it employs IR radiation, radiography, ultrasonic rays, X-rays, and other procedures. Non-destructive testing of a material's thickness, density, and flaws is done by measuring the change in radiation intensity after passing through the material. Non-destructive testing makes use of radioisotopes from sealed sources.

1.2.1 Gamma Imaging

Gamma rays were employed in gamma radiography. The density and atomic number of the stuff passing through matter affect how much less intense gamma radiation is. If there is a crack in a metal casting, the reduction pattern will be different than it would be otherwise. Gamma radiography can therefore be used to inspect welds in a non-destructive manner. For the evaluation of different industrial items with larger thickness, such as boilers, ship, and aeroplane components, and pressure vessels, highly penetrating gamma rays from radioisotope sources like ^{60}Co (1.173 and 1.333 MeV) and ^{192}Ir (average 380 keV) are utilised in radiography procedures. Thermal and nuclear power plants, fertiliser and petrochemical factories, industries for sterilisation, and medical facilities all employ gamma radiography cameras to inspect welds and casts (radiotherapy). Additionally, radiographs are used to investigate the oil flow in hermetically sealed engines and the rate and pattern of material wear.

1.2.2 Computerised Tomography

This system is valuable for obtaining a cross-sectional depiction of the internal forms of test objects. The imaging system consists of a source of gamma rays, a collimated detector, a mechanical manipulator that is precisely controlled, a data acquisition system, and a personal computer. The technique involves gathering transmission data of the penetrating radiation through an item at different planes to create a three-dimensional image. This has already been discussed in previous chapters.

1.2.3 Gamma Scanning

This is a nuclear examination technique commonly used to correct industrial equipment present in refineries and petrochemical facilities, such as reactors and distillation columns. This is carried out without disturbing the ongoing production activity. Process columns are critical parts of extracting and refining raw materials such as petroleum products and fertiliser manufacturing plants. Breakdown of the components of the column would cause inefficiency in the process, leading to the loss of resources or an emergency total shut down of operation. These can lead to remarkable financial losses for the unit and can also affect the economy. To reduce the chances of such adversities, column scanning is done using a gamma ray-emitting radioactive source. A collimated source (containing ^{60}Co and ^{137}Cs) and a detector are placed in the horizontal plane either beyond the diameter (as in tray-type columns) or over the chords (as in packed-bed columns). The source as well as the detector are then advanced synchronously along the stretch of the column, and the intensity of radiation is recorded at the desired heights. The intensity readouts produce the density profile of the apparatus. The investigation of the data with respect to the inner loading and hardware arrangement of the column provides information about the column. "Signature scans" obtained under standard functioning or at the time of pre-commissioning trials are taken for comparison and to obtain useful information about the inner configuration of the column. Gamma scanning gives real-time details and uses a secured and sealed radioactive substance that will not create a radioactive system. It is not disturbed by environmental circumstances and also does not produce any waste in the environment, rendering it safe to make use of gamma scanning methods have widespread applications in correcting bottle necking, preventive maintenance, and optimising of the pattern of industrial columns.

1.2.4 Industrial Tracers

Industrial tracers are applied to distinguish between different substances or items by labelling them, to follow their movements, to record changes in concentration and distribution between phases, etc. Manufacturers use the radiotracers to observe fluid movement and filtration, locate leaks, and measure engine wear and corrosion of the equipment. By adding minute quantities of radioactive substances to the materials used in different processes, it is feasible to study the blending and flow rates of different materials, like powders, liquids, and gases, and to find leaks. In industrial setting, the tracer approach is also used for examining the consistency of mixtures. For example, to examine a chocolate mixture, a minute quantity of short-lived radioisotopes like ^{24}Na or ^{56}Mn is mixed with the primary ingredients. Different samples of the end products are then checked for radioactivity with the help of a GM counter. If each sample shows the same counting rate, that means the blending has been consistent. This technique can also be helpful in mixing processes taking place in the production of cattle food, cement paints, fertilisers, medical tablets, and soap. When testing the sealing process used to create radio valve envelopes, the tracer approach is incredibly sensitive.

In order to analyse processes and identify inefficient factors, radiotracers are also frequently used in industry. They are especially helpful when process modifications can produce noticeable advantages, such as sediment transfer. The oil and gas business also makes use of radioactive tracers to determine the size of potential oil reserves.

Tracer studies are useful for validating design and process data because they allow for the contactless recording of signals without affecting the observed process, have a high detection sensitivity, use a wide variety of tracer nuclides, and may be performed on operational production units.

1.2.5 Inspection

Metal components and the quality of welds are inspected using radioactive elements in many different sectors. The capacity of various types of radiation to permeate materials to varying depths is utilised in commercial gamma radiography. The process is similar to X-rays that are used to screen luggage at airports. However, in gamma radiography, only a tiny pellet of radioactive substance, housed in a hermetically sealed titanium capsule, is required to generate effective gamma rays, as opposed to the huge machine required to produce X-rays. Using this method, even the most delicate parts can be checked for flaws without risking injury. Neutron activation analysis, measuring the humidity level of soil and other materials, detecting explosives in checked luggage, geological bore hole logging, cancer treatment for the human cervix, and other applications are all possible with the neutron-emitting isotope ^{252}Cf. The largest advantage of radioisotopes is that they can be transported to the location where an examination is needed and do not require electricity. The main drawback is that they cannot be turned off, so they must be adequately shielded at all times, not just when in use.

1.2.6 Smoke Detectors

A smoke detector is an equipment that detects smoke, mainly as a signal of a fire. Plastic cases are typically used to contain smoke alarms and detect smoke either optically (photoelectric or optical smoke detector) or by physical process (ionisation detector).

Air atoms are ionised in chambers of ionisation smoke detectors using a tiny radioactive source, ^{241}Am. The electronics present in the smoke detector may detect the resultant micro amperes of current produced from the power grid. When smoke or condensed air enters the ionisation chamber, it disturbs this current because ions adhere to the smoke particles, causing a reduction in the detection circuit. The smoke detector detects the decrease in current generated between the plates and then creates an alert sound. The resulting ionisation current is weighed against the ionisation produced in the sealed compartment. The degree of ionisation depends on the density as well as the atomic number of the medium. Ionisation currents rise as medium density does, for instance, air and smoke. Above a predetermined threshold value, this rise is used to set off an alarm. Numerous public spaces and private homes have these smoke alarms installed. They are so widely used that they account for the majority of radioisotope-based devices in use worldwide. Due to its alpha emission and limited penetration power, ^{241}Am is a suitable source for use in ionisation smoke detectors. This is because it can ionise enough air to create a detectable current while still being safely stopped by the air or the plastic casing of the smoke detector.

A photoelectric smoke detector consists of an incandescent light source, a lens, and a photoelectric receiver (a photodiode). The receiver receives the light after it has travelled through the test air. The circuit detects the intensity of light and sounds an alert if it falls below a fixed threshold, maybe due to smoke. The received intensity of light will be reduced because of scattering from smoke particles, airborne dust, etc.

1.2.7 Use in Mineral Analysis and Processing

The composition of mineral samples is examined using radioisotopes. Nuclear analytical techniques have the potential to increase the efficacy of exploration, extraction, and processing of raw materials, resulting in energy and material savings. These techniques are rapid, specific, sensitive, and simple to use, and the measurements can be used for online process control applications because the results can be obtained in close to real time. This is also helpful in age determination and in the distribution of elements in different rock types. These are also widely used as gamma sources in the continuous assessment of mineral slurry density that controls mineral processing and hydrometallurgical operations.

For oil, gas, mineral, groundwater, or geological exploration, well logging systems use a radioactive source and detecting equipment to locate and record formations deep inside a borehole (or well). Drilling is required in oil and mineral explorations to determine if the ore is worth exploiting; this is an expensive and time-consuming process if thousands of holes are drilled and then all the information is compiled. Numerous nuclear borehole logging devices have been devised that provide direct information about uranium, thorium, and potassium, and thus the mineral composition. This apparatus consists of a shielded detector and a gamma radiation source. The radiation from the source reacts with the neighbouring material. The properties of the radiation are altered in these reactions. The new radiation is measured, and conclusions about the composition of the rock can be derived. Nuclear radiation has the advantage of being extremely penetrating, allowing information to be extracted from boreholes filled with water and from enormous volumes of rock. This method is typically employed in the exploration of hydrocarbons and coal.

The moisture content of coal on conveyor belts can also be determined by gamma ray transmission or scattering. The measurements were made on the basis of the fact that the interactions of gamma rays vary with the atomic number of the substance being bombarded; atoms in water have a distinct atomic number than atoms in coal.

1.3 Radiotracer Applications

1.3.1 Scientific Uses

Numerous scientific and research fields make use of radioisotopes.

- The activity of naturally occurring isotopes in water can be used to estimate the age of underground aquifer water, thereby informing decisions concerning the consumption of groundwater.
- The amount of ^{14}C in the atmosphere roughly doubled as a result of open-air nuclear weapons testing in the 1950s and 1960s. Scientists can make use of the enhanced number of ^{14}C atoms in the atmosphere as tracers to determine soil variation and land degradation. For example, based on the quantity of ^{14}C in organic substances in sediments, scientists can find out whether it was dumped in sedimentary layers before or after the period of weapon testing.
- Levels of particular radioisotopes in environmental samples can be examined to see if countries are abiding by treaties governing the development of nuclear weapons.
- Radioactive materials may be released into the environment as a result of nuclear activity either on a regular basis or as a result of an accident.
- ^{129}Iodine and ^{236}U are traces of nuclear activity, such as the reprocessing of irradiated nuclear material, and can be found in environmental samples.
- Radiotracer techniques are commonly employed to evaluate mathematical models of sediment and pollutant transport in the coastal zone and so contribute to sustainable development.
- Even minute quantities of radioactive material can be readily detected, making it perfect for tracing the movement of water, gases, and even insects.
- To monitor the dispersion of sewage, radioactive tracers that have been combined with sewage are used.
- A analogous method is used to locate small openings in complex systems, such as heat exchangers in power plants.
- With the aid of radioisotopes, it is possible to accurately measure the flow of liquids and gases in conduits, and also the flow rates of large rivers. Due to the sensitivity of radioactivity measurement instruments, radioisotopes are frequently used to detect specific hormones, pharmaceuticals, vitamins, enzymes, proteins, and viruses.

1.3.2 Radiodating

Key Concept
Radio dating was used to determine the approximate age of two major archaeological discoveries: the Shroud of Turin and the Dead Sea Scrolls.

Radiodating is a method of dating rocks and minerals using radioactive isotopes. Through this method, archaeologists can determine when formerly living substances were last alive. This method depends on the existence of ^{14}C, a naturally occurring, long-lived radioisotope found in all living objects. The ratio of ^{14}C to ^{12}C in the environment has been proportionately constant throughout history. If an animal or plant expires, it stops getting carbon, so the quantity of ^{14}C in the tissue starts decreasing via the process of radioactive disintegration. Comparing the ratio of ^{14}C to ^{12}C in the dead with that of the "living ratio" helps to evaluate when the plant or animal last existed. ^{14}C investigation has also been useful during space operations to investigate the possibility of life on Mars. Radioisotope- powered electrical generators have been used to power exploration spacecraft, navigational and weather satellites, and communication satellites. They allow us to operate weather stations at the North and South Poles, seismic sensing stations in remote locations, and devices placed on the ocean floor for scientific investigations and national defence.

Radioactive dating is also used to estimate the age of materials such as glass, pottery, and stone used in ancient times. For long-term dating, isotopes with long half-lives are necessary, such as ^{235}U, ^{238}U, and ^{40}K, whereas for shorter-term dating, ^{14}C is widely used. Detailed applications are already discussed in the previous chapters.

Check Points

- What are the applications of radiation processing?
- What are gauges? Explain their uses.
- What is meant by gamma scanning?
- Write a note on industrial tracers.
- What are smoke detectors?
- Explain different types of smoke detectors.
- What is radiodating?

2 Applications of Radioisotopes in Agriculture

The agricultural and food industries can benefit from gamma ray irradiation in a number of ways:

- Gamma irradiation or neutron irradiation is frequently used along with other techniques to generate new varieties of cereals, root and tuber crops, and oil seed crops.
- Nuclear science technology provides methods for increasing productivity while conserving essential resources for daily life and food packaging. In order to advance agricultural science, radioisotopes have been crucial. Radiation has also been used, for instance, to form new and improved plant varieties from plant embryos.
- Insect control is one more essential application; pest populations are exceedingly reduced and, in certain cases, removed by exposing male insects to sterilising doses of radiation. In the food-growing industry, radiation is most frequently used to help eradicate bugs and bacteria from fruits and vegetables. Insect-related problems can often be resolved without causing any damage to the products involved, either by killing the insects directly or by sterilising the entire male population. Numerous studies may be conducted, such as the development of new fertilisers, nitrogen fixation with legumes, or the development of new and improved pesticides, fungicides, rodenticides, etc. This is not applicable to all insects, but it is very effective against certain varieties of fruit flies. In this fashion, radioisotope radiation is used to kill insects and prevent the loss of agricultural products.
- Plant nutrition research: Radioisotopes are extremely valuable for evaluating the amount of phosphorus and nitrogen accessible in the soil. This estimation aids in determining how much phosphate and nitrogen fertiliser should be added to the soil. Fertilisers tagged with radioisotopes like ^{32}P and ^{15}N have been useful to study their absorption, retention, and utilisation. The excessive use of fertilisers has negative effects on biological diversity and the environment. These isotopes offer a means for determining how much fertiliser a plant absorbs and loses to the environment. Radioisotope tracers enable scientists to increase the efficacy of fertilisers. ^{15}N also aids in the evaluation of nitrogen fixation by plants under field conditions.
- Studying soil properties is crucial to developing efficient farming practices. Radioisotopes can also be used as "tags" to observe the uptake and utilisation of vital nutrients by plants from the soil. This technique helps scientists evaluate the precise nutrient and water needs of crops in specific conditions. Nuclear science and technology have greatly smoothed such investigations and are largely used under soil-plant nutrition experimentation to efficiently use limited water sources. Ionising radiation is also used to cleanse the soil, and a lot of development is going on to use radiation for the elimination of microorganisms in the soil that cause diseases and are damaging to plant life.

2.1 Management of Insect Pests

In plant research, radiation is used to create new plant varieties in order to accelerate the process of creating superior agricultural products. The productivity of agriculture is seriously threatened by insect invaders. They not only lower crop productivity but also spread disease to cultivated commodities.

- Radiolabelled pesticides are useful in monitoring their residues in food, groundwater, soil, and the environment. Such studies have helped identify and reduce pesticide and insecticide adverse effects.
- It has also been used to control mosquitoes and insects all over the world. In addition to strengthening plants, this radiation can be used to reduce insect populations that damage food crops, thereby reducing the use of hazardous pesticides.

IEAE is using nuclear technology to develop environmentally benign alternatives to the Sterile Insect Technique (SIT) for pest control. This method uses ionising radiation to effectively sterilise male insects without impairing their capability to operate in the field or successfully mate with natural female insects. This method entails releasing large numbers of sterilised male insects of the particular species into the field crop.

During sexual reproduction, sterilised male insects compete with the normal male population, and the infertile eggs resulting from their mating produce no progeny. It is a highly special form of “birth control” that decreases and removes the insect population after two or three generations. It has been used successfully to remove several insect pests of agricultural significance around the world.

2.2 Crop Enhancement

Plant breeding necessitates genetic variation of beneficial characteristics for crop improvement. Several forms of radiation are used to cause mutations to create desired mutant lines that are disease proof, higher quality, allow for earlier ripening, and lead to a higher yield. As ionising radiation, for example, gamma and neutron radiation. It offers the possibility of introducing desired traits that were previously absent from nature or had been lost to evolution. An appropriate choice of mutant varieties can increase quality and productivity. By using radiation energy to induce mutation, the Bhabha Atomic Research Centre (BARC) is able to develop several high-yielding varieties of tur, green gramme, groundnut, jute, and rice.

Radioisotopes are also used as a research tool to find new strains of high-quality, drought- and disease-resistant, fast-growing (i.e., ripening), and high-yielding food crops. Radioisotopes also play a role in increasing agricultural output in a sustainable way. These are also extremely helpful for preserving agricultural and dietary products. They are also helpful in the investigation of the optimal utilisation of fertilisers, insecticides, and pesticides on cultivated crops without causing damage to plants and other organisms. Radioisotopes are also used to determine the role of fertilisers in distinct plants.

3 Food Processing and Preservation

The sterilisation of medical equipment and food packaging also makes use of radioisotopes with high gamma ray intensities. Both industrialised and developing nations are experiencing an increase in demand for instant foods that are healthy and have a lengthy shelf life. 25–30% of the world's food production is lost due to microbes and pests before it can be consumed, and these losses are greater in developing countries and are particularly prevalent in hot, humid countries. This loss can be avoided by developing effective food preservation methods. Gamma rays are used to irradiate the food stuffs to destroy microbes in the food that can cause foodborne disease and to control insect and parasite growth in harvested food to prevent wastage and spoilage. The two common sources of high-energy radiation in the food industry are ^{60}Co and ^{137}Cs. Food irradiation with ^{60}Co is often used to prevent sprouting in onions and potatoes, to delay ripening, and to reduce the risk of foodborne illness. ^{137}Cs can be used as a means of enhancing particular properties of various food commodities by means of sterilisation, insect disinfection, delayed senescence and ripening, and sprout inhibition. Gamma radiation from these two isotopes has also been used for the irradiation of many fruits such as papaya, mango, and onion, spices, fish, etc. The extension of the shelf life of certain foods by irradiation is enough to save them from spoiling. Certain seeds and canned foods can be stored for longer periods by gently exposing them to radiation. Food irradiation can replace or drastically reduce the use of food additives and fumigants, which are hazardous for consumers as well as workers in food processing industries. These harmful chemicals are used to eliminate insects from dried fruit, and grains, legumes, and spices. Irradiation does not heat the food material, so food keeps its freshness in its physical state. Radiation also kills the germs without harming the substance that is being disinfected, i.e., without making it radioactive. When treated in this manner, food takes much longer to spoil. This way, cereals, fruits, vegetables, and canned food can be stored for longer periods by gently exposing them to radiation. The agents which cause spoilage (microbes, insects, etc.) are eliminated by irradiation from packaged food, and packaging materials are impermeable to bacteria and insects, so recontamination does not take place. Irradiation of food kills insects and parasites, inactivates bacterial spores and moulds, prevents the reproduction of microbes and insects, inhibits the sprouting of root crops, delays the ripening of fruits, and improves the technological properties of food. Many of the food items are packaged in polyethylene shrink-wrap that has been irradiated so that it can be heated above its usual melting point and wrapped around the foods to provide an airtight protective covering.

The FDA has approved irradiation as a method to inhibit sprouting and delay ripening in many fresh fruits and vegetables. In all parts of the world, there is a growing use of irradiation technology to preserve food. More than 60 countries worldwide have introduced regulations allowing the use of irradiation for food products including spices, grains, fruit, vegetables, and meat.

4 Ecological Studies

Ecologists can benefit from radiotracers as well. For instance, radiotracers can be used to track the migratory and behavioural patterns of a wide variety of species. Ecologists can use radioactive primary producers to investigate food webs by tracking their movement through the food web. Radioisotopes are also used to measure the age of water and the average residence time of waste water (using ^{82}Br). These are also used for determining the direction and flow rate of groundwater. ^{3}H is used to study the dynamics of water movement in different compartments of the hydrosphere. ^{2}H and ^{18}O also provide information about the steady- state characteristics of the water cycle.

5 Nuclear Power Plants

One of the most significant uses of radiation is the production of electricity by nuclear fission. The purpose of a nuclear power facility is to employ a generator that starts with the steam from the boiling water and ultimately generates electricity. In a nuclear power facility, nothing is burned or detonated. Instead, the uranium propellant generates heat by fission. Uranium, which emits radioactive substances, is used to fuel nuclear power facilities. Most of these materials are sequestered within uranium fuel grains or metal fuel rods. However, trace quantities of these radioactive materials (mostly gases) are diluted in the water that is used to cool the reactor. Before being released back into the environment, the water that passes through a reactor is treated and filtered to remove these radioactive impurities. Nonetheless, under controlled and monitored conditions, minute quantities of radioactive gases and liquids are eventually released into the environment.

Check Points

- How are radioactive isotopes useful in agriculture?
- How is the population of insects controlled using radioisotopes?
- What are the advantages of using radiations in food processing and preservation?
- Write a note on nuclear power plant.

Thus, it has been observed that radioisotopes, which decay through the emission of radiation energy, can be utilised in a wide variety of agricultural applications. It would not have been possible to solve many agricultural problems using conventional methods, but nuclear scientists are making steady progress in this direction. In the field of agriculture, radioactive materials are extensively used. Radioactive isotopes have been found to play an important role in several fields, including nutritional studies of trace elements, insect irradiation, soil research, mechanism of photosynthesis, plant genetics, animal research, fertiliser development, nitrogen fixation, plant protection, including insecticide action, metabolism in plants, uptake of fertilisers, ion mobility in soils and plants, food preservation, and plant protection, i.e., insect and pest control. This is due to the fact that these tracer techniques are much easier, more accurate, and more reliable. In addition, radioactive materials and techniques have been utilised in food and agriculture, including soil fertility, plant reproduction and genetics, animal health and production, and food preservation. All of these applications of radiation have improved our way of life. Radioisotopes can also be used to address some environmental issues.

When used appropriately, radioisotopes are a fruitful component of the modern world. The NRC and Agreement States remain committed to safeguarding public health and safety in the use of these nuclear materials by carefully inspecting medical, academic, and industrial applications and monitoring users to ensure safe practices.

6 Radioisotopes and Their Applications

Many of the radioisotopes used by scientists, doctors, and researchers in their work, as well as in consumer products, are also found in the by-products of nuclear power production. Some examples are given in Table 1:

No matter the source, once these radioisotopes have performed their purpose, they all become low-level waste. The only safe way to handle the waste is by disposing of it a permanent low-level radioactive waste facility.

Table 1 List of radioisotopes and their applications

S. No.	Radioisotope	Application
1	^{99m}Tc	Diagnose problems with the brain, bones, liver, spleen, and kidneys
2	^{241}Am	In smoke detectors to protect homes and businesses. To measure levels of toxic lead in dried paint samples; to ensure uniform thickness in rolling processes like steel and paper production; to help determine where oil wells should be drilled
3	^{198}Au	To label sand to study coastal erosion
4	^{14}C	To test drugs so doctors can be sure of no harmful side effects, major research tool; helps in research to ensure that potential new drugs are metabolised without forming harmful by-products; used in biological research, agriculture, pollution control, and archeology; used to measure the age of wood, other carbon- containing materials (up to 20,000 years) and subterranean water (up to 50,000 years)

Table 1 (continued)

S. No.	Radioisotope	Application
5	^{47}Ca	Studying the cellular functions and bone formation in mammals
6	^{109}Cd	To analyse metal alloys for checking stock, scrap sorting
7	^{252}Cf	To inspect airline luggage for hidden explosives; to gauge the moisture content of soil in the road construction and building industries; to measure the moisture of materials stored in soils
8	^{36}Cl	To measure sources of chlorine and the age of water (up to two million years)
9	^{244}Cm	In mining to analyse material excavated from pits; slurries from drilling operations
10	^{57}Co	As a tracer to diagnose pernicious anaemia
11	^{60}Co	In sterilising medical equipment and improving the reliability of industrial oil burners; to sterilise surgical instruments; to improve the safety and reliability of industrial fuel oil burners (industrial radiography); in cancer treatment; food irradiation; gauges; and radiography
12	^{51}Cr	In research in red blood cell survival studies; to label sand to study coastal erosion; also a tracer in the study of blood
13	^{137}Cs	To treat cancer tumours, measure drug dosages, control the flow of oil in pipelines, and check the proper fill level for packaged foods
14	^{67}Cu	When injected with monoclonal antibodies into a cancer patient, helps the antibodies bind to and destroy the tumour
15	^{55}Fe	To analyse electroplating solutions; to detect the presence of sulphur in the air; used in metabolism research
16	^{67}Ga	In medical diagnosis
17	^{3}H	Major tool for biomedical research; used for life science and drug metabolism studies to ensure the safety of potential new drugs; for self-luminous aircraft and commercial exit sign; for luminous dials, gauges and wrist watches; to produce luminous paint, and for geological prospecting and hydrology; to measure 'young' ground water (up to 30 years). Hydrogen-3 in tritiated water; used as a tracer to study sewage and liquid wastes
18	^{123}I	To diagnose thyroid disorders and other metabolic disorders including brain function
19	^{125}I	Major diagnostic tool used in clinical tests and to diagnose thyroid disorders; used in biomedical research
20	^{129}I	To check some radioactivity counters in *in vitro* diagnostic testing laboratories
21	^{131}I	To treat thyroid disorders
22	^{192}Ir	To test the integrity of pipeline welds, boilers and aircraft parts; in brachytherapy/tumour; irradiation; used in gamma- radiography to locate flaws in metal components
23	^{85}Kr	In indicator, lights in appliances such as clothes washers and dryers, stereos, and coffeemakers; used in industrial gauging to gauge the thickness of thin plastics and sheet metal, rubber, textiles and paper; to measure dust and pollutant levels

(continued)

Table 1 (continued)

S. No.	Radioisotope	Application
24	^{54}Mn	To predict the behaviour of heavy metal components in effluents from mining waste water
25	^{24}Na	To locate leaks in industrial pipe line; in oil well studies
26	^{63}Ni	As a voltage regulator, and to detect explosives To detect explosives, and in voltage regulators and current surge protectors in electronic devices; in electron capture detectors for gas chromatographs; used in light sensors in cameras and plasma displays, for long life B-voltaic batteries
27	^{32}P	In molecular biology and genetics research
28	^{33}P	In molecular biology and genetics research
29	^{210}Pb	To date layers of sand and soil up to 80 years
30	^{147}Pm	In electric blanket thermostats; to gauge the thickness of thin plastics, thin sheet metal, rubber, textile, and paper
31	^{210}Po	Reduces the static charge in the production of photographic film and other materials
32	^{238}Pu	To power more than 20 NASA spacecraft since 1972
33	^{238}Pu	Has powered more than 20 NASA spacecraft since 1972
34	^{226}Ra	Makes lighting rods more effective
35	^{35}S	In genetics and molecular biology research
36	^{75}Se	In protein studies in life science research; used in gamma radiography and non-destructive testing
37	^{85}Sr	To study bone formation and metabolism
38	^{90}Sr	In survey meters by schools, the military and emergency management authorities; used in cigarette manufacturing sensors and medical treatment; used for industrial gauging
39	^{198}Au, ^{99m}Tc	To study sewage and liquid waste movements, as well as tracing factory waste causing ocean pollution; to trace sand movement in river beds and ocean floors
40	^{99m}Tc	Most widely used radioactive pharmaceutical for diagnostic studies in nuclear medicine. Different chemical forms are used for brain, bone, liver, spleen, and kidney imaging and also for blood flow studies
41	^{229}Th	Helps fluorescent lights last longer
42	^{201}Tl	In nuclear medicine for nuclear cardiology and tumour detection
43	^{204}Tl	Measures the dust and pollutant levels on filter paper; used for industrial gauging, i.e., gauges the thickness of plastics, sheet metal, rubber, textiles, and paper
44	^{234}U	In dental fixtures like crowns and dentures to provide a natural colour and brightness
45	^{235}U	Fuel for nuclear power plants and naval nuclear propulsion systems; to produce fluorescent glassware, a variety of coloured glazes and wall tiles
46	^{133}Xe	In nuclear medicine for lung ventilation and blood flow studies
47	^{169}Yb	In gamma radiography and non- destructive testing
48	^{65}Zn	To predict the behaviour of heavy metal components in effluents from mining waste water

Further Reading

Amsath A (2008) Textbook of Basic biotechnology (2nd edn.) MMA Publications.

Assimakopoulos PA, Ioannides KG, Pakou AA, Papadopoulou CV (1987) Transport of the radioisotopes iodine-131, cesium-134, and cesium-137 from the fallout following the accident at the Chernobyl nuclear reactor into cheesemaking products. J Dairy Sci 70(7):1338-43. https://doi.org/10.3168/jds.s0022-0302(87)80153-4.

Hussein EM (2003) Handbook on Radiation Probing, Gauging, Imaging and Analysis. Volume II: Applications and Design. Kluwer Academic Publishers Dordrecht Boston London. ISBN 978-90-481-6292-5. ISBN 978-0-306-48403-2 (eBook). https://doi.org/10.1007/978-0-306-48403-2

Mishra UC (2004) Environmental impact of coal industry and thermal power plants in India. Journal of Environmental Radioactivity 72 (1-2): 35-40.

Radioisotopes in Food & Agriculture (2021) Updated by World Nuclear Association.

Sanctis ED, Monti S, Ripani M (2016) Energy from Nuclear Fission. An Introduction. Springer Publications. ISBN: 978-3319306490.

Singh B, Singh J, Kaur A (2013) Applications of radioisotopes in agriculture. International Journal of Biotechnology and Bioengineering Research 4 (3):167-174

The Regulation and Use of Radioisotopes in Today's World (2000) U.S. Nuclear Regulatory Commission. Washington DC. Office of Public Affairs.

Tiwari A, Sachan R, Pandey SR, Kumar M, Bharti P (2022) Radioisotope: Their effect on soil and agriculture. In Advances in Agriculture and Allied Technologies. AkiNik Publications New Delhi.

Dosimetry and Safety Issues

Key Learning Objectives

- Dosimetry and its types
- Types of dosimeters
- Nuclear fallout and its implications
- Radiation exposure and protection
- Government regulations on handling radioactive substances
- Guidelines for prevention of radioactive contamination
- Disposal of radioactive waste

Overview

From the previous chapters, it is now clear that, despite their usefulness in all fields of life, radiation exposure is associated with several risk factors. Radiation dosimetry is a branch of science that explores different methods for the quantitative determination of energy, which is deposited in a given material by ionising radiation, either through direct or indirect exposure. The dosimeters detect and measure the radiation that a person is exposed to and in this way, radiation dosimeters play a role in the safety of a person so that he or she is not exposed to a higher than permissible dose of radiation. Radioactive waste also contributes to the exposure of people, animals, and the whole ecosystem. This radioactive waste may be generated from nuclear reactors, nuclear plants, hospitals, mining, and other facilities. Companies should follow protocols and regulations strictly to ensure that these are disposed of properly and minimise environmental impact.

A. Pathak, *Tools and Techniques in Radiation Biophysics*,
https://doi.org/10.1007/978-981-99-6086-6_14

1 Dosimetry

Dosimetry is the relationship between the energy deposited and absorbed in the attenuating material. This applies both internally, via inhalation or ingestion of radioactive materials (internal dosimetry), and externally, because of exposure to radiation by various sources (external dosimetry). Internal dosimetry assessment is done via a number of monitoring techniques that either measure the radioactive elements found in or excreted by the body. Various instruments and scanners used to detect and measure internal dosimetry have already been discussed in detail in the previous chapters. On the other hand, external dosimetry measurements are done with a dosimeter that measures radiation that can enter the skin, i.e., beta particles, photons, and neutrons, when a certain radiation dose is given to humans. Since beta particles and photons interact via electronic forces, their detection techniques and dosimetry are very different. Here, absorbed energy in matter and in human tissue is measured. Dosimetry is thus widely used for radiation safety and is often used to keep an eye on workers exposed to radiation or unanticipated radiation i.e., in the aftermath of the Three Mile Island, Chernobyl, or Fukushima incidents involving radioactive release. External exposures are monitored using tiny radiation detectors called dosimeters, which are worn by people. The public dosage absorbed is assessed and computed from a number of indications, including ambient gamma radiation readings, monitoring for radioactive particles, and measuring radioactive contamination levels. The information obtained from dosimetry is useful in the estimation of radiation safety, risk analysis, diagnostic dose estimations, and treatment planning. A functional dosimeter has the following desirable properties:

- great precision and accuracy;
- linearity of the signal with dosage over a broad range;
- modest dosage and dose rate dependency;
- little directional dependency;
- homogeneous energy response;
- high speed resolution; and
- a big dynamic range.

1.1 Categories of Dosimetry

There are three general categories of dosimetry:

- *Direct monitoring or personal dosimetry*: This is mainly used to ascertain the doses of individuals who are exposed to radiation as a result of their employment activities.

- *Indirect monitoring:* This method is used to gauge airborne dosage rates when a person resides in a location with a known concentration of airborne radioactivity for a known period of time. This knowledge, along with other information, may be used to compute the radiation dosage a person received during that particular period of exposure. This type of monitoring is used where an airborne radioactive material, such as radon, is the radiation source and when personal monitoring is not possible. This method is also called workplace monitoring.
- *Indirect monitoring using environmental pathways analysis*: In this method, population vulnerability is estimated by calculating the amount of nuclear material in the environment from a source (for example, an exhaust stack) or by monitoring the levels of radioactive chemicals in people's homes. Here doses are calculated using suitable site-specific information such as the concentration of radioactive materials in the atmosphere, water, soil, or locally grown produce.
- *Medical dosimetry*: Here, the required treatment absorbed dose and any collateral absorbed dose are measured. In the medical field, dosimetry is important for dose correlation with clinical results, and sometimes this is used for planning treatment to avoid excess toxicity.

Key Concept
Total effective dose equivalent is a scientific way of determining how doses to a part of the body affect the person as a whole.

2 Dosimeter

It is a tiny radiation detector worn by a person and is used to measure doses from ionising radiation. Dosimeters are classified into two types:

- *Passive dosimeter*: It produces a radiation-induced signal that is stored in the device. The output of the dosimeter is then analysed monthly, quarterly, or semi-annually. This type of dosimeter is worn by the individual, and it provides the most accurate estimation of the individual's exposure to radiation.
- *Active dosimeter:* It generates a signal in response to radiation and provides a live, visual representation of the dose or dose rate that was observed.

2.1 *Types of Dosimeter*

Different types of dosimeters are employed for measuring the dose and dose rate that biological systems are subjected to:

2.1.1 Thermoluminescent Dosimeter (TLD)

The TLD badge contains a sensing material called crystal, which absorbs radiation, and during this process, its structure changes slightly. Actually, some of the electrons are released from the atoms of the crystal. These electrons are held in their excited states by the impurities (also known as contamination centres) inside the dosimeter. To determine how much radiation the crystal has absorbed, the badge is placed in a reading device and rapidly heated. This causes the electrons in the crystal to return to their original state. During this process, visible light is emitted. The intensity of visible light emitted is proportional to the radiation absorbed by the crystal. Thus, a thermoluminescent dosimeter detects ionising radiation. By analysing the brightness of light emitted from a heated crystal in the detector, radiation exposure can be determined. The glow curve depicts the relationship between the quantity of light emitted and the temperature. To determine the dose, this curve is analysed. The crystal in a thermoluminescent dosimeter can be made of lithium fluoride, calcium sulphate, or lithium borate. Once these tools have been "zeroed" after analysis, they are repackaged and reissued for more wear. Some of these gadgets have a long lifespan and are expensive to replace if misplaced. Both the quartz filter and film badge types are being superseded by TLD and electronic personal dosimeters. This instrument measures exposures as low as 1.3×10^{-6} C/kg (5 mR) accurately.

Thermoluminescent dosimeters suffer from a number of limitations. Lithium chips provide accurate spatial information but require careful calibration and readout, which are cumbersome. Also, thermoluminescent dosimeters are oxygen-sensitive, which imposes another limitation. TLDs can only be read once because the data is erased during the read-out procedure.

2.1.2 Optically Stimulated Luminescence (OSL) Dosimeter

This is the most common type of personnel dosimeter used in diagnostic radiology today to monitor occupational radiation exposure. Optically stimulated thermoluminescence (OSL) was first made for radiation therapy but is now also used to diagnose diseases. It contains a carbon-doped aluminium oxide detector and can be read by using laser light at certain frequencies. The working of these dosimeters is quite similar to TLDs; the only difference is that these are passive dosimeters, and instead of heat, a light beam causes luminosity. Optically stimulated luminescence dosimeters assess radiation exposure caused when X rays, gamma rays and beta rays pass radiation through a narrow layer of aluminium oxide. After this, the aluminium oxide responds to laser light by fluorescing at a rate that is directly related to the dose of radiation it has been exposed to. This form of dosimeter is portable, lightweight, and sturdy. Also, these devices are extremely sensitive, have the ability to be re-read, and are more accurate than TLD or film dosimeters. This type of dosimeter can be worn for up to 1 year but is typically worn for only 1 to 3 months before being read out. These instruments deliver an accurate readout as low as 10 Sv for X-ray and gamma ray particles with energies ranging from 5 keV to more than 40 MeV.

However, this dosimeter suffers from certain drawbacks as well: It is more expensive than the Fricke dosimeter (which will be discussed below). It is difficult to prepare and stop being linear at absorbed dosages beyond 10 Gy.

2.1.3 Electronic Personal Dosimeter (EPD)

An electronic personal dosimeter contains three silicon dioxide detectors. Each detector supplies a series of amplifiers and counter circuits to monitor soft gamma, hard gamma, and beta radiations. The outputs from each detector chain are calculated and displayed for deep and skin dosage as well as dose rate. The dosimeter can be reset, typically after obtaining a reading for record keeping purposes, and thus re-used multiple times. EPDs measure in the range between 0.1 mSv and 10 Sv.

Workers who are exposed to ionising radiation on the job are protected by legal restrictions on their annual dose. One of the most widely used methods to determine the amount of exposure an individual receives over a long period of time is to use personal dosimeters. This is another compact electronic gadget with a number of advanced features, including continuous monitoring of cumulative dose and current dose rate, alarm warnings at present levels, and a live readout of the dose accumulated. These are particularly helpful in high-radiation locations, i.e., they provide immediate feedback to responders. If a responder enters a high level radiation area or spends too much time in a low-level radiation area, this dosimeter will alert the responder to the dose rate with an audible alarm, vibration, or both.

Key Concept
The G-value is the number of molecules of reactants consumed or products formed per 100 eV of energy absorbed.

2.1.4 Metal-Oxide-Semiconductor Field-Effect Transistor (MOSFET) Dosimeter

These are unique radiation dosimeters that have extremely small sizes (around 1 mm × 1 mm sensor elements) and allow dose measurements *in vivo* in real time, which is especially important for radiotherapy. There are several advantages to

MOSFET devices. It is a small, portable system. The MOSFET's packed size is less than 4 mm. Because it is based on the MOSFET transistor, the principle of operation is similar to semi- conductor detectors (already discussed). Silicon dioxide serves as the active sensing material in MOSFET dosimeters. Ionising radiation enters the detector and interacts with the semiconductor material and ionises the atoms of semiconductor, producing electron–hole pairs. The incoming radiation creates electron-hole pairs within the silicon dioxide. Electrons swiftly leave out of the gate electrode while holes move towards the Si/SiO_2 interface where they become trapped in long term sites which can persist for years. The electric field created by introducing a negative voltage to the gate is delayed by the rise of holes in the interface area. In order to cause current to flow between the source and the gate, a higher voltage must be provided to the gate. The dose is directly proportional to the change in voltage that occurs before and after exposure. Actually, radiation creates defects in oxide, which in turn affects the threshold voltage of the MOSFET. This change in threshold voltage is proportional to radiation dose. The post-radiation signal is permanently stored and is dose rate independent.

In contrast to TLDs and OSLDs, which can discharge their traps, MOSFET detectors accumulate trap filling, reducing their effective usable lifetime to approximately 10 Gy. They can also function as a persistent dose record thanks to this feature.

2.1.5 Pocket Ionisation Chamber Dosimeter

The pocket ionisation chamber is thought to be the most sensitive form of personnel dosimeter. Ionisation chambers consist of a gas-filled chamber with two electrodes to which a voltage is applied and a conductive outer wall. The wall and collecting electrodes are separated with a high-quality insulator to reduce leakage current when a polarising voltage is applied to the chamber. The current flow that results from the ionisation of air molecules exposed to radiation is measured. They can provide accurate, instantaneous measurements. The chambers are simple to use but lack spatial information. However, this method requires intricate circuitry and is therefore not cost-effective. The efficacy of an ion chamber is dependent on the voltage used for charge collection. These devices are expensive and may provide an inaccurate reading if not checked daily. There are two varieties of pocket dosimeters: self-reading and non-self-reading. This form of radiation dosimeter is rarely used in diagnostic imaging.

2.1.6 Quartz Fibre Dosimeter (QFD)

A quartz fibre dosimeter, also known as a *self-indicating pocket dosimeter* (SIPD) or *self-reading pocket dosimeter* (SPRD), is a pen-like device charged to a high voltage and typically used for a single work cycle only. It consists of an ionisation chamber sensitive to the desired radiation, a quartz fibre electrometer to measure the charge, and a microscope to read the finer image of a scale. It operates by measuring the decrease in electrostatic charge on a metal conductor in an ionisation chamber caused by radiation-induced ionisation of the chamber's air. The quartz fibre is deflected by electrostatic repulsion; the larger the charge, the greater the deflection. A light source projects the deflection of a movable quartz fibre electrode through an objective lens onto a calibrated scale, which is then read through a microscope ocular. It measures the cumulative dose of ionising radiation that the device has received. The device is primarily sensitive to gamma and X-rays, but it can also detect radiation with energies greater than 1 MeV. There are also versions that are neutron sensitive. This dosimeter is typically attached to a person's clothing and worn to measure their actual radiation exposure. The charge exits as the gas in the dosimeter chamber is ionised by radiation, causing the fibre indicator to shift in relation to a graduated scale. An older dosimeter design is the quartz form. It has numerous drawbacks:

1. *Low precision:* Because of the analogue mechanical design, accuracy is approximately 15% less than that of other dosimeters.
2. *Reading errors*: Since it can only be read manually, it is prone to human reading errors.
3. *Small dynamic range:* The charge on the electrode restricts the device's range. The device ceases recording exposures once the battery is depleted.

These are now being superseded by more modern types, such as electronic personal dosimeters.

2.1.7 Film Badge Dosimeter

The film badge dosimeter is a personnel dosimeter used to monitor the cumulative radiation exposure caused by ionising radiation. Film badges are small, portable devices worn by individuals who may be exposed to radiation, such as laboratory workers, X-ray technicians, and nurses. Film badge dosimeters are only to be used once. A monthly-removable piece of photographic film and a holder make up the emblem. An insignia houses the film. Photographic film contains molecules of silver bromide that, when irradiated, form metallic silver. It is a black-and-white photographic film with variable grain size to affect its radiation sensitivity. Film blackening is proportional to the dose. The film is removed from its protective packaging after use and developed to measure exposure. When the film is developed, the optical density is used to calculate the dose that the dosimeter has received during a set time period. Consequently, the level of radiation absorption is indicated by a change in the emulsion of the film, which is visible when the film is developed.

The film badge dosimeter is lightweight and portable and is used to measure and record cumulative radiation exposure from gamma rays, X-rays, and beta particles. Film blackening increases with radiation exposure. To help determine the dose and kind of radiation exposure based on the penetrating capabilities stated above, areas of the badge feature panels made of various materials (aluminium, lead, and plastic). Film badge holders contain multiple filters to determine whether the dose received is for the entire body or for the epidermis. It is typically worn on the coat lapels and as a finger ornament.

Film badge dosimeters apply to situations where real-time information is not needed. It provides a permanent record that is very reliable. However, film badge dosimeters suffer from several disadvantages. These dosimeters cannot be read locally and must be shipped off to be developed; they are single-use only and cannot be refilled; and they cannot be used to measure gamma radiation exposures below 0.2 mSv.

2.1.8 Fricke Dosimeter

This is also called a ferrous sulphate dosimeter. The Fricke chemical dosimeter measures X, gamma and electron exposures based on chemical changes caused by radiation absorption. It is based on the principle of oxidation of ferrous ions (Fe^{2+}) to ferric ions (Fe^{3+}) in aqueous solutions.

$$Fe^{2+} + h\upsilon = Fe^{3+} + e^{-} \tag{1}$$

The increased concentration of Fe^{3+} thus achieved can be determined using a spectrophotometer at 304 nm.

One advantage of the Fricke dosimeter is that it exhibits excellent water and tissue equivalence for dosimetry. The Fricke dosimeter is composed of 96% water by weight, so its dosimeter properties are very similar to those of water. This dosimeter is used for dosage rates up to 106 Gy/s and a dose range of 5 to 40 Gy.

The principal disadvantage of the Fricke dosimeter is its high sensitivity to contaminants, which act as scavengers of the hydroxyl radicals produced by irradiation or as ferrous ion oxidants, resulting in a non-linear response and reduced system sensitivity when the oxygen in the solution is depleted.

2.1.9 Direct Reading Dosimeters

These are active dosimeters since they display dose rates while they are being used and are therefore used in work planning and execution. These dosimeters have alarms that sound at preset doses and dose rates. These have been used as dose control devices in the nuclear power industry.

It is worth mentioning that dosimeters do not protect or shield anyone from radiation exposure. Another important thing is that the dosimeters should be worn consistently in the same place.

2.1.10 Other Methods

Other techniques of dosimetry are also available. They are based on colorimetric measurements, discolouration of crystals, turbidity formation on glasses or organic polymers, the mortality rate of bacteria on exposure to ionising radiation, and conductivity alteration in semiconductors. These methods, however, are not employed so frequently.

Check Points

- What is dosimetry?
- Explain the different categories of dosimetry.
- What is a dosimeter? Explain their different types.
- State the difference between an active and passive dosimeter.

- Thus, radiation dosimeters have important role to play in the fields of health physics and radiation protection. They help in the measurement, calculation and assessment of the ionising radiation dose absorbed by the human body.

3 Safety

The use and disposal of radioisotopes raises many safety concerns. Such issues with radioisotopes can be avoided by using certified laboratories, safety gear, clear directions for use and disposal, proper training, and established international and national regulatory authorities. With low-energy radioisotopes like ^{33}P and ^{35}S which release beta particles with energies of 0.248 and 0.167 MeV, safety is less of a worry. The principal objective should be to incorporate ALARA (as low as reasonably achievable) and environmental protection. Realising that non-radioactive isotopic techniques do not come without their own set of safety issues is crucial. Certain fluorescent and chemiluminescent substrates, as well as their byproducts, are contact allergens and even potential carcinogens.

4 Radioactive Fallout or Nuclear Fallout

Radioactive fallout is the term used to describe radioactive particles that settle to the ground after a nuclear explosion, such as an atomic bombing, a nuclear weapon test, or an accident at a nuclear power station. In the case of a ground burst, these include radiated soil, fission products, and weapon debris. Particles in fallout range in size from a few millimetres to many metres. Thousands of miles after the initial explosion, radioactive particles combine with atmospheric dust and eventually fall to

Earth. The particles that make up radioactive fallout can be as little as the invisible droplets created by an aerosol spray can or as huge as the ash that settles around a wood fire. Within a few minutes of the explosion, much of this debris falls back to the earth, but some ascends high into the atmosphere. These are typically limited to the area downwind of the weapon's burst point. The kind of radioactivity in fallout is determined by the sort of nuclear reaction that released the particles into the environment. Fallout may initially contain more than 60 distinct radioactive compounds. This radiation hazard is caused by radioactive fission fragments with half-lives ranging from seconds to millions of years, as well as soil and other materials near the burst that have been rendered radioactive by the tremendous neutron flux.

Some of these degrade into non-radioactive compounds in seconds, while others take millennia or longer. For instance, it takes 28 years for a sample of ^{90}Sr to lose 50% of its initial radioactivity. One of the most harmful elements in fallout is ^{90}Sr because human metabolism processes it similarly to calcium, a crucial component of bones. A person's bone marrow may be impacted, increasing their risk of developing leukaemia. Food tainted with ^{90}Sr will build up in the bodies of both animals and humans who consume it. ^{134}Cs, ^{137}Cs, and ^{131}I are further dangerous fallout products.

A nuclear explosion can produce local, tropospheric, or stratospheric fallout. The stratosphere, the uppermost layer of the atmosphere, may be exposed to some of the fallout. Fallout must be ejected into the stratosphere by the most intense atomic explosions, such as those created by hydrogen or thermonuclear bombs. Stratospheric fallout can travel for years before it reaches the troposphere and is dumped on the ground. It falls to Earth from the troposphere via electrical or gravitational forces or by attaching to bigger particles like raindrops. Particles from the fallout may cling to vegetation and make their way up food chains to the first consumers. In Lappland (Sweden), where caribou ate toxic grass, fallout from Chernobyl in 1986 was deposited. Shifting winds also brought Chernobyl radiation particles to northern Italy, where rabbit farmers fed their bunnies poisoned plants. The excessive levels of radioactive substances in the rabbits' meat ultimately led to their extinction. The types of radionuclides that enter terrestrial and marine food systems differ. In general, land plants and animals have the largest concentrations of the soluble fission products ^{137}Cs and ^{90}Sr. Marine species are most likely to contain high concentrations of fallout that forms stable compounds with organic matter, such as ^{60}Co, ^{59}Fe, ^{54}Mn, and ^{65}Zn. Colloidal forms of elements like ^{134}Cs and ^{95}Zr are also abundant in marine life. While ^{137}Cs can arise following mishaps at nuclear power plants or as a byproduct of nuclear bomb explosions, ^{134}Cs primarily come from the fission products of a power reactor.

Direct skin contact with fallout particles in the air fallout particles that fell on the ground and then came into contact with the skin eating plants, milk, and meat that had radioactive fallout on or in them, and breathing in radioactive material in the air: are all methods of exposure to radioactive fallout (internal).

5 Nuclear Power Plant Accidents

The International Atomic Energy Agency (IAEA) defines a nuclear or radiation accident as "an occurrence that has led to substantial implications for people, the environment, or the facility". Examples include fatalities among people, radioactive isotopes in the environment, and reactor core melt. The Fukushima Daichi nuclear disaster (2011), the Chernobyl disaster (1986), the Three Mile Island accident (1979), and the SL-I accident (1961) are all examples of serious nuclear power plant mishaps, including fatalities among people, radioactive isotopes in the environment, and reactor core melt. Accidents involving nuclear power can result in fatalities and high repair bills.

6 Radiation Exposure and Radiation Protection

A person's exposure can come from a number of different sources, both external and internal. External exposure can result from direct irradiation from a sealed source or contamination, such as airborne radionuclides deposited on the ground or onto clothing and skin. Internal exposure can result from ingesting contaminated food and water, inhaling radioactive airborne particles, or from contamination of an open wound. Effective defence from external exposure is offered by:

1. *Time*: Limiting the time spent near the source of radiation; for people who are exposed to radiation in addition to natural background radiation.
2. *Distance*: Moving away from the source of radiation will drastically decrease the dose of radiation.
3. *Shielding from the source*: The impact of this depends on the nature of the shielding material and the type of radiation being emitted. From penetrating gamma and X-rays, barriers made of lead, concrete, or water offer protection. This is why dentists cover their patients with a lead blanket during dental X-rays and why certain radioactive items are housed underwater or in concrete or lead-lined rooms.

Thus, time, distance, and shielding actions minimise the radiation exposure.

In contaminated areas, protective clothing and respiratory protection help prevent external body contamination. Eating, drinking, and smoking should be prohibited in contaminated areas. The risk of developing cancer increases with prolonged exposure to low doses of radiation. It can also cause gene mutations, which may be inherited by any children born after the exposure. Radiation exposure for a brief period of time, like during a radiation emergency, can result in burns or radiation sickness. Radiation sickness symptoms include nausea, weakness, hair loss, skin burns, and diminished organ function. If the exposure is high enough, it can lead to premature ageing and even death.

7 Bioaccumulation

More radionuclides have been and continue to be released into the environment as a result of man's increased use of radioactive materials. Once released, they can circulate throughout the biosphere and end up in drinking water, vegetables, grass, meat, etc. The concentration of radionuclides increases as an animal moves up the food chain. This is referred to as *bioaccumulation*. Since many of us eat at the top of the food chain, the process of bioaccumulation of radionuclides can be particularly harmful to humans. Excitation and ionisation occur when radiation interacts with matter or living tissues. A biologist who is exposed to radiation from tracer experiments, which are so common and useful, has a keen interest in radiation exposure, even though everyone in the world is exposed to a small amount of radiation from natural and man-made sources. Therefore, it is beneficial for a biologist to be aware of the tolerable doses and to limit his or her exposure to those levels even further.

Although the "rad" is helpful, the "rem" is a much more useful unit for biological exposure. It considers the varying effects of the studied radiations. In comparison to alpha and gamma radiation, for instance, alpha particles are approximately 10 to 20 times more damaging to tissue. Alpha particles have little penetrating power, so their effects are limited to a small area. Beta and gamma radiation have been assigned a quality factor (qf) of 1, whereas alpha radiation has been assigned a qf of 10. The rem is defined as the dose in rads multiplied by the quality factor.

The annual dose of radiation that the body can tolerate is 5 rem (hence 0.1 rem per week). The more delicate tissues, such as the gonads and bone marrow, were removed to create this total body dose (because the fast dividing tissues are more sensitive to radiation). There are, however, other tissues that are less sensitive. Thus, the hands and forearms, which are most exposed when conducting biological experiments, can tolerate a higher exposure. The permissible exposure limit for these organs is 75 rem per year and up to 20 rem in any 13-week period.

8 Government Regulations on Radioactive Substances

Each government has specific regulations for the secure use of radioactive substances, and each institution that wishes to use radioactive substances must obtain a copy of these regulations and submit an application for a licence to import and work with radioactive materials.

1. Some of the duties of the licence holder dealing with the use of radioisotopes are mentioned below:

    (a) Provide suitable housing for the laboratory for radiation containment.
    (b) Appoint a radiation officer and an assistant radiation officer.
    (c) Create internal guidelines for handling radioactive compounds.

   (d) Create contingency plans and estimate the risks.
   (e) Be in charge of supervised areas.
   (f) Observe stock levels.
   (g) Ensure that equipment is maintained and that methods are operating optimally.
   (h) Ensure that monitoring equipment is calibrated.
   (i) Keep an eye on contamination and radiation levels.
   (j) Ensure that everyone who works with radioactivity is properly trained and informed of the dangers.
   (k) Give personnel dose-measuring equipment and keep track of doses received.
   (l) Limit the amount of radiation exposure that employees are exposed to.
   (m) Notify authorities of any radioactive substance loss or release.
   (n) Inform authorities and the affected worker of overexposure if it occurs, and investigate and report how it transpired.

2. Radiation officers must have experience regulating all radiation-related activities. Radiation officers are responsible for ensuring compliance with national regulations and serve as the institution's point of contact with the licensing body (government, etc.).
3. Internal regulations and documents should be drafted and reviewed annually. The following documents must be readily available at all times: the government regulations pertaining to radioactivity, the licence, and the internal rules. Copies of these documents must be provided upon request to any worker.
4. Controlled areas are typically separate rooms from the primary laboratory. The entrance to the room should be restricted to trained personnel only, and the doors should be marked with warning signs. For areas where radioactive materials may be present in limited quantities, such as a laboratory with a gel documentation system for gel analysis with radiolabelled PCR, a secondary classification of supervised areas may be used.
5. Stock records must include the substance's name and activity, the date it was acquired, whether it was sealed or not, what it was used for, when it was used, and how it was disposed of. The radiation officer should inspect the record on a regular basis. All records must be kept for 5 years following the last entry.
6. Regular baseline level monitoring in radiation areas and laboratories is required to ensure that maximum dose limits are not exceeded. Monitoring devices must be calibrated upon purchase and after repair.
7. Records of monitoring and calibration must be maintained.
8. Radiation must be contained during storage, and warning signs must be present. Radioisotopes must be stored in secure environments (locked rooms, sealed freezers, or refrigerators). Internal norms must include disposal guidelines.
9. Radiation employees must wear a dose-measuring device. Dosimeters must be replaced on a monthly basis or after an exposure is suspected. Dose limits must not be exceeded, and any overexposure must be reported to the licencing authority.

10. Employers must inform employees of the loss or emission of radioactive substances and equipment defects.
11. The department should have warning signs, shielding, personal protection equipment, ventilation, safety mechanisms, and warning devices to limit radiation exposures.

9 Basic Principles for Handling Radioactive Material in the Laboratory

These can be summarised as:

- Planning and practice.
- Time.
- Distance.
- Shielding and monitoring.
- Containment.

Planning and practice: Before radioactivity is introduced to the location, everything needed for the experiment should be assembled. Prior to conducting a new procedure involving radioisotopes, the procedure should be tested using non-radioactive material.

Time: The dose from the radiation source can be decreased by limiting or minimising exposure time. Every effort should be made to limit the time spent directly handling radioactive compounds. Speed must be encouraged, but not to the point where mistakes are made. Planning and practice are the two fundamental principles used to reduce exposure time. In order to expedite the procedure and/or reduce the amount of time the patient is exposed to radiation, it is necessary to plan the procedure in advance and evaluate alternative methods of performing the task prior to the actual procedure.

Distance: The intensity of the radiation drops as one moves away from the source. It decreases proportionally to the square of the distance, so doubling the distance between a labourer and the source reduces the exposure rate by a factor of four. Consequently, some protection can be gained by ensuring that the radioactive source is as far away from the body as possible. Using remote handling instruments, distance can be maximised. Other employees should leave the area when handling highly radioactive isotopes. If possible, storage areas must be located away from common work areas.

Shielding: Shielding is the most important method of protection when working with radioisotopes. In practice, the majority of laboratory work employs beta emitters either low-energy (^{3}H, ^{35}S) or high-energy emitters (^{32}P). Gamma and X ray-emitting nuclides like ^{51}Cr, ^{125}I, ^{131}I, and ^{99m}Tc and necessitate specialised screens and detectors. However, shielding is not strictly necessary when beta particles (^{14}C, ^{3}H, and ^{35}S) are released because they have low energy and do not propagate far in the air. Beta particles, however, are highly energetic and dangerous because of their penetrating power. Therefore, protection from this radioisotope is essential. To reduce the quantity of bremsstrahlung radiation produced, the shielding material must be of low density (e.g., acrylic or Perspex) and not metal. ^{32}P is shipped from the manufacturer in special containers (often made of lead and plastic) that minimise exposure; these containers should be kept for stockpiling purposes. The original vial in which the isotope is purchased, the microfuge tube in which reactions are conducted, and the refuse container (an empty plastic chemical bottle or wastebasket) should suffice.

When managing low-energy beta emitters, it is crucial to avoid skin and laboratory surface contamination as well as internal exposure (e.g., ingestion or contamination through a cut). Accidental ingestion of the radioisotope can cause the majority of the dangers associated with ^{14}C, ^{3}H, and ^{35}S. External exposure to radiation emitted by these nuclides is not a concern unless they are present as contamination on the skin's surface.

Containment: Radioactive materials should be kept out of the environment and contained in the smallest practicable space. Dispensing of radioactive materials should take place in sealed handling facilities to prevent contamination of the larger laboratory.

10 Guidelines for Prevention of Radioactive Contamination

- Contaminated work gloves are the main source of unintentional contamination of laboratory surfaces. Whenever direct handling of isotope open containers is done, it should be presumed that gloves have become contaminated. Disposable mittens should be changed frequently because tritium can penetrate them. Using the GM survey metre, the mittens should be checked for isotope content on a regular basis.
- The pipetting of radioisotopes is another typical source of contamination. It occurs when the pipette tip releases aerosols, drips, or flying particles. Plugged tips should always be used to prevent aerosol production. In addition, after the sample has been pipetted out, the tip of the microfuge tube should not be removed until the plunger has been released and permitted to return to a neutral position.
- Reduce the length of time the bare, potentially contaminated tip is on the work area to lessen the consequences of drips. This is accomplished by properly positioning and handling the waste and sample containers.
- When the contaminated pipette tip is not removed neatly from the tube but is instead pulled up the side, flying droplets are produced. If this occurs, the tip will provide a little "flick" as it enters the tube. It not only causes the tube's mouth to become soiled, but it also causes droplets to spray out. Tips must always be removed from tubes without touching the walls.
- Lack of preparation is a significant contributor to contamination. Before actually carrying out an action, it is a good idea to mentally practise it.

Despite all precautions, contamination may occur occasionally. It is therefore essential to operate in a manner that prevents the subsequent spread of that contamination. Continuous monitoring is the only way to prevent unwittingly spreading contamination. The Geiger counter should be activated before beginning a radioactive experiment. Throughout the investigation, it must be on. During the radioactive experiment, employees should monitor themselves with a Geiger counter. It is preferable to inspect hands and handheld instruments after each action. Sleeves and arms are also "hot areas" for contamination. Other parts of the body should be monitored less strictly, but at least at the conclusion of the experiment. Before leaving the radioactive location, shoes should be examined. It must be guaranteed that the Geiger counter is suitable for this radioisotope.

- One should be aware of the regulations governing the use of radioactivity. It takes a series of errors to pose a problem. If the rules are correctly applied, one error (such as a leak) will not be too dangerous because the other rules will compensate for it (such as the wearing of gloves, the quick discovery of the spillage by monitoring, etc.).
- Any laboratory or facility that becomes contaminated should be decontaminated by the authorised user. In addition, upon vacating all premises where radioactive materials have been used, the authorised user is responsible for removing and disposing of all residual radioactive materials. Before leaving areas where radioactive materials have been used, the Radiation Safety Office must be notified. Decontamination refers to the process of removing radioactive materials from a person, object, or aircraft. Use specialised detergents for radiation decontamination, such as Decon 90 and Contrad 70. These should be diluted according to the manufacturer's recommendations for surface decontamination (typically a 10–20% solution in deionised water). If these are not available, dishwashing detergent can be used, but it may be less effective at removing contamination. It is advantageous to use a detergent bottle that can be operated with one hand and is of the "squeezable" variety. Never start labour without tissues and detergent on hand. Decon 90 and comparable detergents may erode the surface of contaminated metal items (but in any case, avoid metal due to bremsstrahlung X-ray formation).
- The prenatal exposure limits will go into effect once the radiation worker has informed the Radiation Safety Officer that she is pregnant and has declared it in writing. Declared pregnant workers (DPW) may be given two badges: one for the entire body, worn on the upper torso, and one for the unborn child, worn on the lower belly.

11 Radioactive Waste Management

Key concept
The International Commission on Radiological Protection is an international advisory body that provides recommendations and guidance on radiation protection.

This involves two stages: collection and disposal. The radioactive waste should be identified and segregated within the area of work according to the different classifications discussed above. Each package should be monitored and labelled for activity level before deciding on the mode of disposal. Combustible radioactive waste should be segregated from non-radioactive waste, and this combustible waste material should be disposed of through incineration.

11.1 Collection and Storage of Radioactive Waste

In the lab, each isotope should be kept in its own clearly labelled trash can. This is due to the fact that various techniques are used to dispose of them. Even radioisotopes that can be disposed of together (such as ^{32}P and ^{35}S) should be separated before being placed in the waste container. This is due to the fact that low-energy isotopes can be handled more readily, and the two isotopes require different levels of shielding. For the same reason, it is preferable to separate high-activity trash from low-activity trash wherever possible. There are two tiers of garbage collection. The first is the trash, which is gathered in trash cans on the countertop. The bags should be moved to a more permanent storage facility once they are full or if radiation work is to be suspended for an extended length of time. Transferring something dangerous requires extra caution.

11.2 Disposal of Radioactive Waste

The safe disposal of radioactive waste is crucial. The primary goal of radioactive waste management is to keep radiation exposure to humans and the environment below acceptable levels. A number of laws and regulations strictly govern the disposal of radioactive waste. The Atomic Energy (Safe Disposal of Radioactive Waste) Rules of 1987, which were issued by the Indian Central Government Atomic Act 1962, govern the disposal of radioactive waste in the public domain in India.

Radioactive waste is classified in a number of ways:

11.2.1 According to the Level of Activity

- *High-level waste*: This is a highly radioactive byproduct of nuclear reactor processes. This includes either the spent (used) reactor fuel when it is accepted for disposal or the waste products that remain after the spent fuel is reprocessed. Due to the presence of highly radioactive fission products, this sort of waste requires extremely long-term storage and disposal.
- *Medium-level waste*: This type of waste contains long-lived radioisotopes in concentrations similar to high- level waste and therefore also requires isolation and containment for periods greater than several 100 years. This type of waste includes refurbishment waste, ion exchange resins, and some radioactive sources used in radiation therapy.
- *Low-level waste*: Items that have been polluted with radioactive substances or rendered radioactive by exposure to neutron radiation fall under this category of garbage. This kind of garbage includes used medical tubes, swabs, injection needles, syringes, contaminated protective shoe covers and apparel, wiping rags, mops, filters, and laboratory animal corpses and tissues. Licensees keep low-level waste on-site until it can be disposed of with regular garbage or until there is enough of it to warrant shipping to a low-level waste disposal site in "containers approved by the Department of Transportation".

11.2.2 According to Form

- *Sanitary Sewer*: The Institute may release licensed radioactive materials into the sanitary sewerage system provided that:
  - The material is readily soluble in water or is a readily dispersible biological material; and
  - The annual total of all radioactive materials released into the sanitary sewerage does not exceed 1 Ci (37 GBq).
- *Solid Waste*: Nuclear power reactors produce solid waste as a result of processing contaminated liquid and gaseous effluent as well as other solid materials that have gotten contaminated through unintentional or intentional contact with radioactive materials. Solid radioactive waste should typically be picked up from the laboratories on a fixed day only, and the trash must be readied for collection by:
  - Dividing the trash into two piles, one for short-lived trash and the other for long-lived trash.
  - Separating the liquid scintillation vials and sharps from the solid trash
  - Separation of radioactive and non-radioactive waste is required. Under no circumstances may radioactive waste be thrown away with regular garbage.

- *Liquid Waste*: The use of radioisotopes in industry and hospitals results in the production of liquid waste during nuclear reactor operations. The type of operation that was carried out will determine the chemical make-up and radioactivity levels of the wastes that are produced. Liquid radioactive waste is often gathered at the same time as solid radioactive waste. Toluene, xylene, and other commonly collected liquid radioactive wastes are toxic to fish, shellfish, and the environment since they are not water-soluble. If the liquid waste is not water soluble, it should be stored in a plastic container with a label indicating the contents and the activity. The Radiation Safety Office can provide you with the plastic storage containers.

11.2.3 According to Half-Life

- *Long-half-life waste:* This contains radioisotopes with a half-life of more than a month.
- *Short half-life waste*: This contains radioisotopes with a half- life of less than a month.

11.3 Methods of Radioactive Waste Disposal

The collected radioactive waste is disposed of in the following manner:

- *Dilute and disperse*: Low-activity solid articles may be disposed of as ordinary waste if the activity of the article does not exceed 1.3 μCi or the overall package concentration does not exceed 135 $\mu Ci/m^3$. Examples include vials, syringes, cotton swabs, tissue papers, etc. Similarly, liquid radioactive waste with less activity can be disposed off into the sanitary sewer system with sufficient water flushing following the disposal.
- *Delay and decay*: It is possible to store radioactive waste with medium activity and half-lives of less than a month. An exhaust system is required to properly ventilate the storage space. In order to prevent radiation leakage, the storage area should have lead shielding with a suitable thickness (10 HVL). When barely 0.1% of the original activity is left after decay, the radioactive waste should be stored for at least 10 half-lives. The waste is then checked for residual activity, and if the dose limit is met, it is disposed of as low-activity solid and liquid waste.
- *Concentrate and contain*: This radioactive waste disposal method is used for radioactive materials with very high activity levels and extended half-lives (longer than a month). Due to the lengthy storage period required, it is a common practice to dispose of such isotopes using the delay and decay method. Therefore, radioactive material is collected in containers that are appropriately designed and labelled and then buried in exclusive burial sites approved by the competent authority.

- *Incineration*: This is used to dispose of insoluble liquid waste, for example, from the liquid scintillation systems. The majority of the waste is reduced via incineration, and the activity is concentrated in a smaller volume of ash for disposal, such as solid radioactive waste. However, incinerators used to dispose of radioactive waste discharge a portion of the radioactivity into the atmosphere. They should therefore operate under regulated conditions and in isolated locations.

Check Points
- What are the safety issues related to the disposal of radioisotopes?
- State the problems associated with nuclear fallout.
- What are internal and external radiation exposures?
- How can the radiation exposure be minimised?
- State the government's regulations while working with radioactive substances.
- Write down the principles of handling radioactive material in a laboratory.
- How can radioactive contamination be prevented?
- On what basis are radioactive isotopes segregated?
- Write down the methods of disposal of radioactive waste.

Thus, it can be concluded that the dose of various ionising radiation and the distribution of the dose absorbed by a patient can be determined by radiation dosimetry techniques using various types of dosimeters. Here, the radiation doses are estimated and assigned to the individuals. Dosimeter badges can track exposure over several weeks or months, while pocket dosimeters are more suitable for short-term monitoring. Radiation safety is in charge of making sure that all clinical and scientific uses of radioactive materials are carried out in a way that is safe for all parties involved, including staff, patients, and the general public. These days, radioisotopes are increasingly being used for a number of applications. All these will result in a rise in radioactive waste that must be appropriately disposed of in accordance with regulations set by natural organisations like India's Atomic Energy Regulatory Board (AERB). Lastly, the government guidelines providing information on the safe handling of radioactive substances should be strictly followed to make the best use of radioisotopes in day-to-day life.

Further Reading

Abaza AMH (2018) New trend in radiation dosimeters. American Journal of Modern Physics 7(1): 21-30

Al Hameed HMS (2019) Study on the application of phototransferred thermoluminescence to reassessment of radiation dose using the MCP-N and MTS-N detectors. PhD thesis. Department of Nuclear Physics and Radiation Safety. University of Lodz

Baldev R, Venkataramen B (2004) Practical Radiography. 162–163. UK: Alpha Science Int'l. ISBN 1-84265-188-9

Burn RR (1988) Introduction to Nuclear Reactor Operation

Choudri BS, Charabi Y, Baawain M, Ahmed M (2017) Radioactive wastes. Water Environ Res 89(10):1487-1502. https://doi.org/10.2175/106143017X15023776270539

Ciancaglioni I, Consorti R, Rossi MC, Conte G (2004) Polycrystalline diamond detectors compared with silicon X-ray dosimeters for clinical use. IEEE Symposium Conference Record Nuclear Science 2004. https://doi.org/10.1109/NSSMIC.2004.1466871

deAlmeida CE, Ochoa R, de Lima MC, David MG, Pires EJ, Peixoto JG, Salata C, Bernal MA (2014) A feasibility study of Fricke dosimetry as an absorbed dose to water standard for 192Ir HDR sources. Plos One 1–13. https://doi.org/10.1371/journal.pone.0115155

Glossary of electric and natural gas industry terms and concepts (2022) Indiana Utility Regulatory Commission

Grupen C, Rodgers M (2016) Radioactivity and Radiation. What They Are, What They Do, and How to Harness Them. Springer Publications

Joiner M (2021) Radiobiology of high dose per fraction. International Conference on Advances in Radiation Oncology. Vienna, Austria. Organised by International Atomic energy Agency Feb 16-19, 2021

Keepin GR (1965) Physics of Nuclear Kinetics (1st edn.) Addison- Wesley Pub Co

Khan S, Syed AT, Ahmad R, Rather TA, Ajaz M, Jan FA (2010) Radioactive waste management in a hospital. Int J Health Sci (Qassim) 4(1): 39–46

Khandpur RS (2019) Compendium of Biomedical Instrumentation. John Wiley and Sons Ltd Print ISBN:9781119288121. https://doi.org/10.1002/9781119288190

Knoll GF (2010) Radiation Detection and Measurement (4th edn.) Wiley ISBN-13: 978-0470131480

Lamarsh JR (1972) Introduction to Nuclear Reactor Theory (2nd) Addison-Wesley, Reading, MA

Lamarsh JR, Baratta AJ (2001) Introduction to Nuclear Engineering (3rd edn.) Prentice-Hall. ISBN: 0-201-82498-1

Mallick S, Rath GK, Benson R (2020) Practical Radiation Oncology. Springer Publications

Martin JE (2013) Phydose of various ionising radiations and the sics for Radiation Protection (3rd end.) Wiley-VCH. ISBN-13: 978-3527411764

Naeem AS (2007) Physics and Engineering of Radiation Detection. 647–648. USA: Academic Press. ISBN 978-0-12-045581-2.

Organization of a Radioisotope Based Molecular Biology Laboratory (2006) International Atomic Energy Agency

Paul F (2021a) Pocket chambers and pocket dosimeters. ORAU Museum of Radiation and Radioactivity. Oak Ridge Associated Universities

Paul F (2021b) Robley Evan's lauritzen electroscope. ORAU Museum of Radiation and Radioactivity. Oak Ridge Associated Universities

Pickler A, Mantuano A, Salata C, Mota CL, David MG,de Amorim GJ, Magalhães LAG, deAlmeida CE (2021) Analysis of the Fricke solution-PMMA interaction and potential effect on the accuracy of dose measurements. Radiation Measurements 140: 106513. https://doi.org/10.1016/j.radmeas.2020.106513

Reuss P (2008) Neutron Physics. EDP Sciences. ISBN: 978-27598-414

Sesonske G (1994) Nuclear Reactor Engineering: Reactor Systems Engineering (4th edn.) Springer. ISBN: 978-0412985317

Stabin MG (2010) Radiation Protection and Dosimetry: An Introduction to Health Physics, Springer. ISBN-13:978-1441923912

Stacey WM (2001) Nuclear Reactor Physics. Wiley & Sons. ISBN: 0-471-39127-1

Starter R, Howarth D (2012) Gamma radiation. In: Gamma Radiation. Adrovic F (ed) IntactOpen. https://doi.org/10.5772/34856

Tandon P, Prakash D, Kheruka SC, Bhat NN, Srinivasan VSS, Pandya A (2011) Dosimetry aspects of hafnium oxide metal-oxide-semiconductor (MOS) capacitor. Thin Solid Films 520: 574-577

Upadhyay A, Upadhyay K, Nath N (2022) Biophysical Chemistry (Principles and Techniques) 3rd Edn. Himalaya Publishing House pet. Ltd., Mumbai. ISBN Number: 978-93-5142-227-3

U.S. Department of Energy, Instrumentation and Control. DOE Fundamentals Handbook (1992) Volume 2

Williams WSC (1991) Nuclear and Particle Physics. Clarendon Press: 1 edition. ISBN: 978- 0198520467

U.S. department of energy. Nuclear Physics and Reactor Theory. DOE Fundamental Handbook Volume 1 and 2

Glossary

Accelerator An accelerator is a machine that can accelerates charged particles to speeds close to the speed of light.

Active dosimeter This dosimeter provides a direct display of the accumulated dose and also has some additional functions of alarm threshold settings for dose or dose rate values.

Adjuvant therapy This is the additional cancer treatment administered after the initial treatment to reduce the likelihood that the cancer will return.

Air kerma Air kerma is defined as the number of individual X-ray photons per unit area.

ALARA ALARA stands for as low as reasonably achievable. It means averting exposure to radiation that provides no direct benefit, even if the dose is low.

Alpha decay It is a type of nuclear decay during which helium nucleus is emitted.

Alpha particles Alpha particles are high-energy positively charged particles consisting of two protons and two neutrons.

Amplified fragment length polymorphism It is a PCR-based method that employs the selective amplification of a subset of digested DNA fragments to generate and compare fingerprints for genomes of interest.

Anger type scintillation camera It is a device that is used to image gamma-emitting isotopes through a technique called scintigraphy. These gamma photons are captured by this camera and are converted into light and then to a voltage signal.

Annihilation It is the condition of being completely destroyed or obliterated.

Anthropogenic radiations These radiations are generated by the chemical, industrial or nuclear activities, and result in the release of large amount of radioactivity in the environment.

Anti electron Also called positron. It is an antimatter counterpart of electron.

Apposition autoradiography This is also called temporary contact method of autoradiography. It is mainly used for large specimens.

Atom It is the smallest unit of a substance that cannot be chemically decomposed.

A. Pathak, *Tools and Techniques in Radiation Biophysics*,
https://doi.org/10.1007/978-981-99-6086-6

Atomic number It is the number of protons present in the nucleus of an atom.

Auger electrons Auger electrons are extremely low-energy electrons emitted by radionuclides that decay via electron capture.

Autoradiography Autoradiography is an imaging technique that is used for the localisation of radiation emitted from radionuclides contained within the exposed sample by using photographic emulsions.

Average life It is the average lifetime of all nuclei of a specific unstable atomic species.

Background These are the radiations that all humans are exposed to even in the absence of visible radioactive sources.

Becquerel The Becquerel is the unit to measure radioactivity. One Becquerel is equal to one disintegration per second,

Beta decay It is a type of radioactive decay in which a proton is transformed into a neutron and vice versa within the nucleus, releasing a beta particle.

Beta minus decay In this type of decay, a neutron is converted into a proton, electron, and antineutrino.

Beta particle This is a high-energy, high-speed particle (electron or positron) that is emitted from the nucleus during beta decay.

Binding energy The minimum quantity of energy required to remove an electron from a particular shell in an atom is known as the electron's binding energy

Bioaccumulation Bioaccumulation refers to the progressive buildup of substances within an organism.

Biological half-life The time elapsed before half of a compound (whether radioactive or not) has been removed by biological means is referred to as the biological half-life.

Boron neutron capture therapy In this therapy, boron-containing substance is infused into a blood vessel. Boron accumulates in tumour cells. Neutrons are then utilised to administer radiation therapy to the patient. Neutrons interact with the boron to eliminate tumour cells without injuring healthy cells.

Breakdown voltage It is the threshold voltage at which breakdown is initiated.

Bremsstrahlung radiation Also called braking radiation. It is the electromagnetic radiation emitted by free electrons that are deflected (i.e., accelerated) in the electric fields of charged particles and atomic nuclei.

Carrier free radionuclide This is a radioactive isotope preparation that is "free" of stable isotopes of the element in question.

Characteristic X-ray When outer-shell electrons occupy a vacancy in the inner shell of an atom, X-rays that are released are called characteristic X-rays and they have a "characteristic" pattern for each element.

Chemical quenching If a substance in the sample interfere with the transfer of energy from the solvent to the primary flour or from the primary flour to the secondary flour, the process is called chemical quenching.

Cold spot A cold spot is an area that does not absorb any radionuclides.

Collimator A collimator is a fine processing layer of a gamma camera to receive photons for radioactive source.

Colour quenching Colour quenching occurs if the sample is coloured. This results in some of the light being absorbed before it leaves the sample vial.

Compton effect During Compton effect, the medium energy incident gamma photon delivers a portion of the energy of the electron, which gets ejected from an atom.

Continuous discharge During continuous discharge, radiation is not required to generate discharge because the voltage is so high. The strong electric field ejects the electrons from atomic shells; atoms and molecules become ionised, and a discharge can occur even in the absence of radiation.

Cosmic rays Cosmic rays are high-energy particles or clusters of particles that travel nearly at the speed of light through space.

Cosmogenic radionuclide When cosmic rays interact with gases and particulate matter in the Earth's atmosphere, cosmogenic radionuclides are produced.

Counting efficiency Counting efficiency is the proportion of emission events that generate a detectable photon pulse.

CT scan A CT scan uses X-rays to show a cross-section of the body.

Curie Curie is defined as the amount of radioactive material in which the rate of disintegration per second is identical to that of radium.

Cyclotron A cyclotron is a form of particle accelerator that propels a beam of charged particles (protons) in a circular path repeatedly.

Dead time Dead time is the time interval following a pulse during which the counter becomes insensitive to subsequent ionising events.

Decay constant It is the fraction of the isotope disintegrating in a unit time.

Decay energy The decay energy is the change in nuclear energy caused due to radioactive decay.

Decay series or chain The series of daughter nuclides formed by the radioactive decay of a parent nuclide to a stable daughter nuclide constitute the decay chain.

Diagnostic radiopharmaceuticals These are the agents that are used to diagnose certain medical problems.

Diagnostic radiology Diagnostic radiology is the branch of medicine that uses ionising radiation for the diagnosis of disease.

Digital radiology Digital radiography is a form of X-ray inspection technique that generates an instantaneous digital radiographic image on a computer.

Directly ionising radiations These are the electrically charged particles with enough kinetic energy to produce ionisation through collision.

Disintegration It is the process by which an unstable atomic nucleus dissipates energy via radiation.

DNA foot printing DNA foot printing is a technique used to identify protein-binding regions in DNA molecules.

Dose calibrator A dose calibrator is an instrument that is used to measure the activity of radionuclide samples.

Dose equivalent It is a unit that quantifies the biological effectiveness of an absorbed dose of ionising radiation.

Dosimeter It is a tiny radiation detector worn by a person and is used to measure doses from ionising radiation.

Dosimetry Dosimetry is the relationship between the energy deposited and absorbed in the attenuating material.

Dot blot assay This is a probe based rapid screening molecular biology technique that facilitates detection of proteins in different samples.

Double isotope study It is a study that is used to measure two different isotopes in one sample by using a scintillation counter.

Dynode A dynode is an electrode within a vacuum tube that functions as an electron multiplier via secondary emission.

Effective dose The effective dose is a quantity that considers the reaction of different organs and tissues in the body to the radiation. This quantity is mainly used during radiation protection.

Effective half-life It is the amount of time required for the activity of a radioisotope deposited in a living organism to be reduced by 50 percent as a result of physical decay and biological elimination.

Electromagnetic force It is the force exerted on a charged particle as it moves through an electric and magnetic field.

Electron An electron is a negatively charged sub-atomic particle.

Electron capture In this type of decay, one of the electrons from the inner orbital is attracted into the nucleus and is therefore captured by the nucleus. It integrates with the protons to form a neutron; thus, there is a loss of one proton and a gain of one neutron per nucleus.

Electronic personal dosimeter The electronic personal dosimeter is a modern electronic dosimeter used to estimate the dose of ionising radiation absorbed by the individual wearing it.

Emission computed tomography Emission computed tomography refers to the functional imaging technique that combines the tracer kinetic method and tomographic image reconstruction.

Equivalent dose Equivalent dose compares the effectiveness of different types of radiation on biological tissues.

Excitation The transition of the electrons to the higher orbitals on absorption of energy is called excitation.

Excited state Excited state is the common name for the higher shells.

External dosimetry External dosimetry is the precise measurement of personnel radiation exposure outside the body.

External radiotherapy External radiotherapy is a type of local treatment that uses machinery to direct high-energy radiations from outside the body to the inside of diseased cancer cells and to shrink tumours, and therefore can treat extensive regions of the body.

Fast neutron therapy Fast neutron therapy is a highly effective and unique form of external beam radiation therapy that is used to treat specific malignancies that are radio resistant or exceedingly challenging to eradicate with standard X-ray radiation therapy.

Fission Fission is the splitting of an atom into two or more smaller ones caused by the collision of a neutron with a larger atom.

Fixatives Fixatives are the chemicals that are used to take off any remaining unbound silver halide in the emulsion.

Fluence It is defined as the radiation energy delivered to a given region per unit time.

Fluor Fluor is a mixture of compounds that absorb excitation energy from the solvent and produce some fraction of this energy as light.

Fluorescence It is the phenomenon of immediate emission of light of longer wavelength.

Focussing grid It is a part in the photomultiplier tube that is used to collect the electrons emitted from the photocathode or dynodes efficiently and produce an electrical signal.

Fricke dosimeter Fricke dosimeter, also called ferrous sulphate dosimeter, is one of the most useful chemical dosimeters in existence.

Gamma camera A gamma camera, also called a scintillation camera, is the most commonly used imaging device in nuclear medicine.

Gamma decay In this type of decay some unstable atomic nuclei dissipate excess energy through a spontaneous electromagnetic process.

Gamma imaging It is the process of utilising the penetrative properties of photons of gamma radiation emitted by the nuclei of radioactive elements to create diagnostic images.

Gamma knife Gamma Knife is an advanced radiation therapy technique that makes use of computerised treatment planning software to help physicians precisely locate and irradiate small targets within the cranium and brain. The treatment delivers high doses of radiation to the target area while sparing the adjacent tissue.

Gamma rays Gamma rays are high-energy particles that can pass completely through the human body, causing ionisations that can damage tissue and DNA.

Gantry It is a structure that supports the detector head (consisting of a collimator, scintillation crystal, PM tubes, and amplifiers) and allows the detector head to be moved to the appropriate position for imaging a patient.

Gas amplification Gas amplification occurs when the initial ions produced by the reaction of a photon with a gas molecule cause additional ionisation reactions, which multiply the output pulse measured across the detector.

Gas-filled detector It is based on the ionisation of gas molecules by radiation, followed by the collection of the ion pairs as charge or current when a voltage is applied between two electrodes.

GM plateau It is the area in the GM curve where further increase in voltage results in a negligible increase in counting rate.

GM tube The GM tube is a hollow cylinder filled with a gas at low pressure. The tube has a thin window made of mica at one end through which radiation enter the tube.

Gravitational force It is the force that draws objects towards the centre of a planet or other body.

Gray The gray is the SI unit of absorbed dose and is defined as the absorption of one joule of energy.

Ground state It is the stationary state of lowest energy.

G-value The G-value is the number of molecules of reactants consumed or products formed per 100 eV of energy absorbed.

Half-life It is the time required for a quantity (of substance) to reduce to half of its initial value.

Half value layer It is the distance needed to reduce the number of gamma rays by half.

Heteroduplex tracking assay It detects sequence differences between normal DNA and the DNA being tested.

High level waste This is a highly radioactive byproduct of nuclear reactor processes.

High dose rate brachytherapy It is a form of internal radiation where a cancer doctor implants a small plastic tube or balloon (a catheter) in the tumour; places highly radioactive material inside the body for a short time; and then takes it out using a remote control.

Hot spot A hot spot is the area which has a greater amount of radionuclides.

Image guided radiation therapy It is a type of radiation therapy that allows for taking photographs of a patient throughout the course of treatment, i.e., before the beginning of the treatment and also during radiation therapy treatment.

Immunoradiometric assay This assay uses radiolabelled antibodies in place of antigens. The antibodies are tagged with radioisotopes that bind the antigens in the specimen. It is capable of providing higher sensitivity than RIA.

Incineration Incineration is the process of burning hazardous substances at high enough temperatures to eradicate contaminants. This is used to dispose of insoluble liquid waste.

Indirectly ionising radiation They themselves do not cause any chemical or biological damage but release secondary electrons (or charged particles) after absorbing energy in the material.

***In situ* hybridisation** It is a technique that allows for precise localisation of a specific segment of nucleic acid within a histologic section.

Intraoperative radiation therapy Intraoperative radiation therapy refers to radiation treatment administered during surgery.

Intensity-modulated radiation therapy This is a more advanced type of 3-D CRT that targets the tumour from several directions and avoids healthy tissue.

Internal conversion Internal conversion is a non-radiative process, as no photons are emitted in the process.

Internal dosimetry Internal dosimetry is the science and discipline of determining the internal ionising radiation dose due to radionuclides within the human body.

Internal radiotherapy Internal radiotherapy means having radiation treatment from inside the body.

Interstitial radiation therapy With interstitial radiation therapy, the implants are inserted into or near the tumour location, such as within the breast or prostate, but not in a body cavity with the help of wires, balloons, and small seeds the size of rice grains are used using CT scans, ultrasounds, etc.

Intracavity radiotherapy In this type of radiation therapy, the source containing radioactivity is placed inside a body cavity (space), like the rectum or uterus, using a tube or cylinder that fits in a particular body opening.

***In vivo* foot printing** *In vivo* foot printing enables the monitoring of alterations in DNA-protein binding that occur within a living cell.

Ionisation It is the process of formation of ion pairs after absorption of energy.

Ionisation chamber Ionisation chamber is a form of radiation detection apparatus. Two opposing electrodes are placed in a gas-filled container in an ionisation chamber, and a high voltage is applied. As charged particles (radiation) pass through a gas, the molecules of the gas are ionised, producing ions and electrons.

Ionising radiation Ionising radiation is a type of energy that removes electrons from the atoms and molecules of substances, including air, water, and biological tissue.

Ion survey meter This is a gas-filled detector mainly used to determine the presence and intensity of radiation.

Isobar Isobars are the isotopes of various elements with the same mass number.

Isobaric transition During this transition, nucleons transform into each other to obtain a more stable configuration with the emission or absorption of a beta particle.

Isomeric state It is a state of an atomic nucleus having a different energy and observable half- life from those of other states of the same nucleus.

Isomeric transition It is a nuclear process in which a nucleus has abundant energy following the emission of an alpha or a beta particle and in turn discharges energy without a change in its number of protons or neutrons.

Isotones Isotones are nuclides containing an identical number of neutrons.

Isotopes Isotopes are the atoms of the same element having same atomic number but different mass numbers.

K capture Also called electron capture. It is a process in which the captured electron usually comes from the atom's K-shell.

Latent image A latent image is an invisible image created by exposing a photosensitive material, such as photographic film, to sunlight.

Linear accelerator/linac It is a machine used to produce X-rays or photons.

Linear energy transfer It is the rate at which electron energy is transferred to a medium and is usually expressed as the quantity of energy delivered per unit of distance.

Liquid emulsion It is a mixture of two or more liquids in which one is present as droplets, of microscopic or ultramicroscopic size, distributed throughout the other.

Liquid waste Liquid waste consists of liquid that are the hazardous or potentially harmful to human health or the environment.

Lower level discriminator It is a device used for processing analogue electrical pulses generated from a photon detector. Pulses are transmitted only if they correspond to the events of energy greater than a certain threshold.

Low-level waste Low-level waste consists of materials contaminated with radioactive material or rendered radioactive by neutron radiation exposure.

Man made or technogenic radiation This type of radiation is generated in the range of medical, commercial, and industrial activities.

Mass deficit When a nucleus is formed, some of the mass is converted to energy and this results in the mass deficit.

Mass number It is the sum of the number of protons and the number of neutrons in the nucleus.

Medical radiation It is a type of radiation that is used in many medical imaging procedures.

Medical resonance imaging Magnetic resonance imaging is a type of scan that employs powerful magnetic fields and radio waves to produce detailed images of the body's interior.

Medium level waste This type of waste contains long-lived radioisotopes and includes refurbishment waste, ion exchange resins, and some radioactive sources used in radiation therapy.

Meson It is a subatomic particle which is intermediate in mass between an electron and a proton and transmits the strong interaction that binds nucleons together in the atomic nucleus.

Metal-oxide-semiconductor field-effect transistor dosimeter These are unique radiation dosimeters that have extremely small sizes (around 1mm x 1mm sensor elements) and allow dose measurements *in vivo* in real time, which is especially important for radiotherapy.

Metastable state It is the excited state of an atom, nucleus, or other system that has a longer lifetime than the ordinary excited states.

Microarrays Microarrays is a collection of high-density microscopic single stranded DNA attached to a solid surface.

Multi-channel analyser It analyses a stream of voltage pulses and sorts them into a "spectrum" of number of events, versus pulse height, which may often relate to energy or time of arrival.

Minimal erythema dose It is the reddening of the skin exposed to the shortest duration of UV radiation.

Muon A muon is a type of particle very much like an electron-exactly the same as an electron—except heavier.

Natural radiation Natural radiation are those radiation that are produced from cosmic rays, and terrestrial sources, such as radionuclides in the earth's crust, air, food, and water, as well as the human body itself,

Neo-adjuvant therapy Neo-adjuvant therapy is given before the main treatment, to help reduce the size of a tumour or kill cancer cells that have spread.

Neutrino Neutrinos are tiny, virtually massless particles that travel close to the speed of light.

Neutron Neutrons are the subatomic particles found inside the nucleus of every atom.

Non-ionising radiation It is a type of low-energy radiation that does not have enough energy to remove an electron from an atom or molecule.

Novel targeted therapy This type of therapy is more specific in inhibiting the growth of cancer cells and may improve the efficacy of other treatments.

NORM Naturally occurring radioactive material belong to NORM and contains a material found in the environment that contains radioactive elements of natural origin such as radium, uranium, thorium, and potassium and their radioactive decay products.

N-type material An N-type material is a semiconducting material used in electronics that is produced by introducing an impurity into a purified semiconductor, such as silicon or germanium. Phosphorus, arsenic, antimony, bismuth, or another chemical element may be used as impurities.

Nuclear explosion It is an explosion caused by the rapid discharge of energy from the detonation of a nuclear device. Nuclear detonation can cause significant destruction and loss of life due to blast, heat, and radiation.

Nuclear fallout Nuclear fallout is the radioactive residue ejected into the upper atmosphere by a nuclear explosion.

Nuclear imaging Nuclear imaging is a procedure that uses gamma-emitting radioisotopes for creating an image.

Nuclear medicine Nuclear medicine employs radioactive material within the body to examine organ or tissue function (for diagnosis) or to target and eliminate damaged organs or tissues (for treatment).

Nuclear reactor A nuclear reactor is an apparatus used to initiate and control fission or fusion nuclear chain reactions.

Nucleus The nucleus is the small, dense region consisting of protons and neutrons at the centre of an atom.

Nucleons This is a collective name given to protons and neutrons that are present in the nucleus.

Nuclide It is a distinct kind of atom or nucleus characterised by a specific number of protons and neutrons.

Operating voltage It is the voltage at which equipment is being operated.

Optical quenching It is the attenuation of the photons of light.

Optically stimulated luminescence dosimeters This is the most common type of personnel dosimeter used in diagnostic radiology today to monitor occupational radiation exposure.

P-10 gas P 10 gas is a mixture of Argon 90% + methane 10%.

Pair production It is the production of a particle—antiparticle pair from the decay of a neutral particle or from a pulse of electromagnetic energy travelling through matter, usually in the vicinity of an atomic nucleus.

Palliative care/palliation It is a type of therapy that helps to decrease the symptoms due to developing tumours and enhance the quality of life.

Palliative radiation Palliative radiation is the use of radiation to cure a specific area, and in some cases to alleviate side effects like pain and bleeding.

Parallel hole collimator The parallel hole collimator is the most widely used multi-hole collimator in nuclear medicine laboratories. It consists of parallel holes with a long axis perpendicular to the plane of the scintillation crystal.

Particle accelerator A particle accelerator is a machine that produces a stream of charged, high-velocity atomic or subatomic particles and smashes them into a target.

Particulate radiation These radiations consist of particles having mass as well as energy and may or may not have an electrical charge.

Passive dosimeter A passive dosimeter is a device that do not provide direct readouts and can operate without any active means.

Peptide receptor radionuclide therapy It is a treatment that targets some types of neuroendocrine tumours. It binds to specific receptors on the tumour cells and destroys them with a small, but powerful dose of radioactivity.

Penetration power It is the ability which allows an electron to get close to the nucleus is known as penetration.

Permanent brachytherapy In this technique, the implants are permanently placed in or close to the tumour using a delivery device (either ultrasound or X-ray imaging) to ensure their correct placement.

Phosphor It is a fluorescent material that produce tiny flashes of light when it absorbs radiation.

Photoelectric cathode It is coated with a photosensitive material that causes the release of electrons when photons pass through it.

Photoelectric effect The photoelectric effect refers to the interaction of a low-energy γ-ray photon with the electrons in the inner shell of an atom of the material and completely transfer their whole energy to the inner electrons, which get ejected as a photoelectron from the atom.

Photographic developers Photographic developers help to indicate the silver grains as a darkening of the film.

Photographic emulsion Photographic emulsion is a fine suspension of insoluble light-sensitive crystals in a colloid sol, usually consisting of gelatin. The light-sensitive component is one or a mixture of silver halides silver bromide, chloride, and iodide.

Photoluminescence detector It uses a photoluminescent material that absorbs and stores photons ("particles" of light) from a light source and on removal, the stored energy is released as visible light and "glows in the dark".

Photomultiplier noise The main source of photomultiplier noise is thermionic emission of the electrons from the photocathode, and from the first few dynodes of the electron multiplier.

Photon It is a minute energy packet of electromagnetic radiation.

Pin hole The image in the pin hole camera is created on the basis of the rectilinear propagation of light.

Pixel Pixels are the smallest unit in a digital display.

Pocket ionisation chamber dosimeter It is used to provide the wearer with an immediate reading of his or her exposure to X-rays and gamma rays

Positive holes It is a vacancy which is created at the site of a covalent bond when an electron leaves a covalent bond.

Primary ion pair Ion pairs are created when an atom or molecule undergoes the process of ionisation.

Proton beam therapy It is a type of radiation therapy that uses protons to painlessly delivers radiation to treat some types of cancer.

Photo multiplier tube It is a light-sensitive device that converts light released into measurable electronic pulses.

Polymerase chain reaction Polymerase chain reaction (abbreviated is a laboratory as PCR) is a technique for rapidly producing (amplifying) millions to billions of copies of a specific segment of DNA.

POPOP 1,4-(di-2-(5-phenyl oxazolyl))-benzoyl (POPOP) is used as a secondary scintillator.

Positive beta decay In this type of decay, a proton in the parent nucleus decays into a neutron that remains in the daughter nucleus, and the nucleus emits a neutrino and a positron, which is a positive particle like an ordinary electron in mass but of opposite charge.

PPO 2,5-biphenyl oxazole (PPO) is used as a primary scintillator.

Positron Positron is a positively charged electron-like particle.

Proportional counter It is an instrument used for detecting and measuring the intensity of ionising radiation. It operates at a lower potential difference such that the magnitude of the discharge is directly proportional to the number of gas molecules ionised by the detected particle.

Proton Proton is a fundamental positively charged particle with the same amount of positive charge as electrons.

Primary cosmic ray Primary cosmic rays (mainly protons) are stable and charged particles that have been advanced by astrophysical sources situated in the universe.

Primary fluor It is the fluorescent substance to which radiation energy is directly transferred from excited molecules of the solvent in liquid scintillation counter.

Primordial radionuclide This type of radionuclide originates from earth long before its creation; these are actually the remnants of the Big Bang cosmogenic sources and ancient supernova explosions that took place before the solar system was formed.

P type material P type material contains gallium and indium as impurity that is added to a pure semiconductor in a small amount, and as a result, a large number of holes are created in it.

Pulse height analyser A pulse height analyser is a device for analysing the frequency distribution of the spectrum of photon energies captured by a gamma camera.

Pyrosequencing Pyrosequencing is a method of DNA sequencing that relies on the detection of pyrophosphate release and the generation of light on nucleotide incorporation.

Quantitative foot printing This is used to estimate the protein's binding affinity using varying concentrations of the protein for the foot printing experiment and observing the appearance of the footprint as the concentration is increased.

Quartz fibre dosimeter It is a pen-like device charged to a high voltage and typically used for a single work cycle only.

Quenching Quenching is the decrease in counts caused by the properties of the sample or cocktail and its components.

Quenching agent It absorbs the emitted energy which dissociates the agent while preventing further ionisation.

Radiation absorbed dose It is defined as the amount of ionising radiation absorbed by the irradiated tissues.

Radiation biology This is the branch that is concerned with how radiation affects and interacts with living things and systems.

Radiation dose It is the amount of radiation absorbed into one's body.

Radiation exposure Radiation exposure is a measure of the ionisation of air due to ionising radiation from photons.

Radiation weighting factor The radiation weighting factor is a dimensionless factor used to calculate the equivalent dose from the absorbed amount of radiation.

Radiation oncology This is a kind of radiation therapy that is used to treat cancer (using high-energy X-rays or other particles).

Radioactive decay Radioactive decay involves a transition from the original nuclide's (parent) quantum state to the product nuclide's (daughter) quantum state.

Radioactive decay law The law of radioactive decay states that the number of radioactive nuclei decaying per unit time is directly proportional to the total number of nuclei in the sample material.

Radioactive element Radioactive elements are composed of atoms with unstable nuclei that emit atomic radiation as a means of achieving stability.

Radioactive fallout Radioactive fallout is the term used to describe radioactive particles that settle to the ground after a nuclear explosion, such as an atomic bombing, a nuclear weapon test, or an accident at a nuclear power station.

Radioactive series/chain In radioactive series, unstable heavy atomic nuclei decays through a sequence of alpha and beta decays until a stable nucleus is achieved.

Radioactivity Radioactivity is the property of spontaneous disintegration possessed by certain unstable nuclei (radionuclides), and this disintegration is always followed by the release of high-energy radiation that includes alpha or beta particles or gamma rays.

Radio dating Radio dating is a method of dating rocks and minerals using radioactive isotopes.

Radio ecological half-life It is defined as the radioactive half-life of the animals and vegetation residing in a particular area. Radio ecological half-lives are even less accurate than physical and biological half-lives.

Radioimmunoassay A radioimmunoassay is an immunoassay that employs radiolabelled molecules to form immune complexes in a stepwise manner.

Radiometric dating It is a method for dating materials such as minerals and carbon in which trace radioactive impurities were selectively incorporated during their formation.

Radioimmunotherapy Radioimmunotherapy is a combination of radiation therapy and immunotherapy in which a radioactive substance is attached to a monoclonal antibody and injected into the body.

Radioisotope Radioisotopes are the isotopes that are unstable, or radioactive, and give off radiation spontaneously.

Radionuclides Radioactive forms of elements are called radionuclides.

Radiopharmaceutical A radiopharmaceutical is a drug that contains a radioisotope bound to a biological molecule.

Radiopharmaceutical therapy In this type, the radiopharmaceuticals are administrated intravenously to be delivered to a target tumour. Once a radiopharmaceutical has stuck to a cancer cell, the radioactive compound naturally breaks down and releases energy that damages the DNA of nearby cells.

Radio protectors Radio protectors are substances that protect healthy tissues near the treatment area.

Radio resistant It is the level of ionising radiation that organisms are able to withstand.

Radiotherapy Also called radiation therapy in which the high-energy radiation is used to permanently damage the DNA of cancer cells, causing them to die.

Recombination region In the recombination region, the voltage is low such that ion pairs generated by a ray or a particle (called primary ion pairs) are moving so slowly that they can easily recombine to form neutral atoms or molecules, i.e., the electrodes only capture some of the primary ion pairs.

Relative biological equivalent It is the ratio between the dose (in rads) of a radiation that is to be applied and the dose of γ-rays producing the same biological effect.

Restriction fragment length polymorphism In restriction fragment length polymorphism analysis (RFLP), specific restriction endonucleases recognise and cleave DNA molecules at certain places, resulting in fragments of the DNA.

Rutherford It is the unit that measures how many nuclei decay per second in radioactive material.

Riboprobe A riboprobe, also called RNA probe, is a segment of labelled RNA that can be used to detect a target mRNA or DNA during *in situ* hybridisation.

Roentgen The roentgen is the amount of X-rays or gamma rays (less than 3 MeV) such that the related corpuscular emission per 0.001293 g of air at STP produces in air ions carrying 2.58×10^{-4} coulombs (1 esu) of the quantity of electric charge of either sign.

Roentgen equivalent man It is defined as the dose that is delivered to a man who is exposed to any type of ionising radiation, i.e., the amount of energy deposited by the radiation in body tissues or organs, when human risk factors are present.

Roentgen equivalent physical Roentgen equivalent physical (rep) is defined as the dose of any ionising radiation that produces an absorption of 83 ergs of energy per cubic cm of tissue.

Sanitary sewer A sanitary sewer is an underground conduit or tunnel system that transports sewage from residential and commercial buildings (but not stormwater) to a sewage treatment facility or disposal facility.

Saturation voltage It is a broad voltage range in which the response is uniform, so that even if the voltage varies, a similar number of electrons will be gathered.

Scalar It consists of an electronic counter that is used for counting the pulses coming from an amplifier or SCA either for a specific time interval or until a predetermined number of counts have been collected. In either case, both the number of counts and the time during which they are collected are obtained.

Scintillation It is a physical process where a material, called a scintillator, emits UV or visible light when excited with high-energy photons (X-rays or γ-rays) or energetic particles (such as electrons, alpha particles, neutrons or ions).

Scintillation cocktail Scintillation cocktail consists of a solvent, that dissolves the sample, and fluor, that emits light.

Scintillation crystal It is a light-sensitive device that absorbs gamma photons and converts them into visible light.

Scintillator It is a substance that emits fluorescent light when exposed to gamma ray photons.

Secondary cosmic ray It is produced as a result of the disintegration of primary cosmic rays as they interact with the atmosphere.

Secondary electrons These are created as a result of inelastic interactions between the primary electron beam and the sample.

Secondary fluor A second fluor absorbs the fluorescent light emitted by a primary fluor and is thus excited to emit light at a higher wavelength, which may be more convenient for detection.

Secondary radionuclides These are obtained from the disintegration of primordial radionuclides.

Seed Seed is a type of implant—smaller than a grain of rice—that is used in radiation therapy that delivers a high dose of radiation directly to the target organ.

Self-indicating pocket dosimeter This is another name for quartz fibre dosimeter.

Sievert It is the International System of Units (SI) derived unit of dose equivalent radiation that takes into account the relative biological effectiveness of different forms of ionising radiation.

Self-reading pocket dosimeter This is another name for quartz fibre dosimeter.

Simulation Simulation is the use of CT, MRI, and PET scans to plan the treatment area.

Single channel analyser It is a device used for measuring pulse amplitude in the nuclear field.

Single photon emission computed tomography scan It is a nuclear medicine technique which helps in the diagnosis and monitoring of diseases.

Single stranded conformational polymorphism It is a simple and sensitive technique for mutation detection and genotyping.

Smoke detector A smoke detector is an equipment that detects smoke, mainly as a signal of a fire.

Soft emitter The radioisotope that has low energy and a short penetration distance is called soft emitter.

Solid-state detector Solid-state detectors generate a signal by accumulating the charge emitted when a particle passes through a semiconductor.

Solid waste Solid waste is produced by nuclear power reactors as a result of processing contaminated liquid and gaseous effluent as well as other solid materials that have gotten contaminated through unintentional or intentional contact with radioactive materials.

Spallation Spallation is a violent reaction in which extremely high-energy particles bombard a target.

Specific activity Specific activity is defined as the amount of a given radionuclide's activity per unit of atoms.

Specific gamma ray constant It is the rate of exposure per unit of activity at a specific distance from the radiation source.

Specific ionisation Specific ionisation is the number of ion pairs produced per unit track length.

Stereostatic radiation therapy Stereostatic radiation therapy uses high-energy focused beams to treat a very small area of the tumour. This is mainly practised when the tumour cannot be safely handled with surgery.

Stripping film emulsion This film is available while fixed to a support. It is removed from the support and directly used on the sample.

Strong force Strong force is a force that holds a nucleus together and acts between proton—proton, proton—neutron, and neutron—neutron.

Synchrotron A synchrotron is a cyclotron in which the magnetic field intensity increases with the particle's energy to maintain a constant orbital radius. It is a potent X-ray source.

Systemic radiation therapy In systemic radiation therapy, the drugs containing radionuclides are given orally or injected intravenously. These drugs then travel through the blood throughout the body, locating and killing tumour cells.

Targeted alpha therapy Targeted alpha therapy represents an emerging treatment approach and provides the opportunity to bypass mechanisms of acquired resistance in selected tumours.

Tele therapy This is another name for external radiation therapy.

Temporary brachytherapy In temporary brachytherapy, the radioactive material is placed inside the catheter for a specific period of time, and then it is removed.

Terrestrial radiation Terrestrial radiation is produced by radioactive elements that were present in the earth's crust five billion years ago.

Therapeutic application This includes the treatment and care of a patient with the goals of preventing and combating disease, as well as relieving suffering or injury.

Therapeutic nuclear medicine Therapeutic nuclear medicine is developing as an alternative choice of treatment in oncology and pain management.

Therapeutic radiopharmaceutical Therapeutic radiopharmaceuticals are used to treat a previously diagnosed disease or disorder.

Thermoluminescent detector Thermoluminescent detectors are used in radiation monitoring. When heated, the electrons return to their ground state and light is emitted

Thermoluminescent dosimeter Thermoluminescent dosimeter measures the absorbed dose from all types of ionising radiation.

Three-dimensional conformational therapy It is one of the most common types of external radiation therapy. This therapy uses the images obtained from CT,

MRI, and PET scans to obtain detailed three-dimensional images of the treatment area.

Threshold voltage Threshold voltage of the counter is the voltage where the counter starts counting.

Tomogram This is high-resolution three-dimensional images produced by gamma camera.

Tomotherapy Tomotherapy is a form of IMRT that employs a device that combines the capabilities of a CT scanner and the external- beam radiation machine

Total effective dose equivalent Total effective dose equivalent is a scientific way of determining how doses to a part of the body affect the person as a whole.

Townsend avalanche effect This is an ionisation process for gases in which free electrons are accelerated by an electric field, collide with gas molecules, and release additional free electrons. In turn, these electrons accelerate and liberate additional electrons.

Upper level discriminator It is a device for processing analogue electrical pulses generated from a photon detector.

Weak force Weak forces play a significant role in nuclear transformation. These forces govern the beta decay process and the interactions of neutrinos with nuclei.

X-ray These are the type of electromagnetic radiations that are used to create pictures of the inside of the body.

X-ray film X-ray film displays the radiographic image and is composed of a single or double emulsion of silver halide that, when exposed to light, produces a silver ion (Ag^+) and an electron.

Index

A. Pathak, *Tools and Techniques in Radiation Biophysics*,
https://doi.org/10.1007/978-981-99-6086-6

T

MIX
Papier aus verantwortungsvollen Quellen
Paper from responsible sources
FSC® C105338

If you have any concerns about our products,
you can contact us on
ProductSafety@springernature.com

In case Publisher is established outside the EU,
the EU authorized representative is:
Springer Nature Customer Service Center GmbH
Europaplatz 3, 69115 Heidelberg, Germany

Printed by Libri Plureos GmbH
in Hamburg, Germany